Implementing a Mixed Model Kanban System

The Lean Replenishment Technique for Pull Production

Implementing a Mixed Model Kanban System

The Lean Replenishment Technique for Pull Production

"An implementation workbook for your next challenge on the lean journey...when you cannot achieve continuous flow."

James C. Vatalaro & Robert E. Taylor

PRODUCTIVITY PRESS • NEW YORK

Most Productivity Press books are available at quantity discounts when purchased in bulk. For more information contact our Customer Service Department (888-319-5852). Address all other inquiries to:

Productivity Press
444 Park Avenue South, 7th floor
New York, NY 10016
United States of America
Telephone: 212-686-5900
Fax: 212-686-5411
E-mail: info@productivitypress.com

Cover design by Gary Ragaglia
Page design and composition by William H. Brunson, Typography Services
Printed and bound by Malloy Lithographing in the United States of America

Library of Congress Cataloging-in-Publication Data

Vatalaro, James C.
    Implementing a mixed model Kanban system : the lean replenishment technique for pull production / James C. Vatalaro & Robert E. Taylor
        p.   cm.
    Includes bibliographical references and index.
    ISBN 1-56327-286-5
    1. Just-in-time systems.   I. Taylor, Robert E.   II. Title.
TS157.4.V38   2003
658.5'1—dc21

                                                                2003006070

"Think of inventory as a powerful narcotic. When prescribed and used correctly, it could save your life . . . when abused, it could lead to your demise. Are you an abuser?"

JCV

CONTENTS

PREFACE

Many books have been written about controlling production and replenishment systems with "lean" approaches, such as continuous or one-piece flow and cellular manufacturing. In fact, these approaches are the best way to add value to your products *in the areas of your value stream that permit it*. If you are very fortunate, the approaches characterized by flow can be deployed throughout your entire value stream. However, this is very rarely the case. We have worked in a broad spectrum of companies around the globe and have discovered a common reality. In most areas of the value stream, implementing effective one-piece flow is either not physically possible or not the logical thing to do given the existing process technology.

We are painfully aware of the typical "current state" situation. The customer wants the product delivered in smaller lot sizes and delivered more frequently, in less time, at a lower cost, with better quality and better customer service. The suppliers are trying hard to respond by controlling production and/or replenishment using the traditional "solutions." These solutions include costly (both financially and emotionally) computer hardware and software, "hot" or expedite lists, and the old "whoever shouts the loudest" method. In some cases you may have attempted creative solutions that exhibit some of the attributes of a kanban pull system, but at the end of the day you find yourselves wondering, "Did we make the customer happy by making the right things?" If this describes your operation, you will find *Implementing a Mixed Model Kanban System* of great benefit.

Few of the tools of the Toyota Production System, also known as "just-in-time production" or "lean manufacturing" are as misunderstood and undocumented as kanban. Kanban is a term very frequently referred to today, but a deep understanding of it has been elusive. Therefore, a clear and concise step-by-step kanban implementation manual is long overdue. The objective of this workbook is to remove the mystery and provide a reliable implementation process.

Implementing a Mixed Model Kanban System begins where the vast majority of the previously published material on value stream mapping and continuous flow production finishes, and it explores new territory. We assume that the user is familiar with the basic concepts of both. If not, a wealth of information on these two prerequisite subjects is available through Productivity Press.

So, what exactly should you do if you fit the low-volume, high-mix mold, and you have many shared supplying resources, batch processes, or production equipment that is nearly impossible to dedicate and co-locate into a continuous flow "cell"? The answer is actually very simple: you have three choices. The first is to hire a qualified consultant to tell you where in your value stream you should implement kanban, how to perform a requirements study, how to determine order frequency, how to calculate the number of kanban needed, how "the system" will work, and how it must be managed on a daily basis. The second is to use the straightforward, low-tech, and flexible approach found in this manual as a step-by-step guide in the implementation of kanban in your value stream. The third is to use a combination of the two.

The approach we describe in this workbook is not theoretical; it is based on a proven, reliable method that has been honed over years of implementation experience. This kanban implementation experience is not just limited to the manufacturing environment. Kanban has been successfully deployed in many nonmanufacturing environments where reliable replenishment is just as important. Some examples include retail, government procurement, hospitals, and administrative processes. With very little adaptation to suit your environment, all the forms and worksheets necessary for you to implement kanban pull in your company, or as we say "Go-Do," are included in both hard copy and electronic format.

Your willingness to "Go-Do" is one of the main reasons why we decided to write *Implementing a Mixed Model Kanban System*. We recognize that you have been constrained by the absence of two things. First, a lack of educational material on the principles of kanban pull systems. Second, by not having a step-by-step process to follow to implement it. Your constraints are about to be eliminated. To "Go-Do" immediately best sums up our challenge to you.

We wish you great success in your kanban pull implementation efforts.

James C. Vatalaro & Robert E. Taylor

ACKNOWLEDGMENTS

The authors would like to gratefully acknowledge the contributions of the following individuals:

Joseph P. Cardinale Jr.

Paul McGrath

Jacqueline O'Banks Vatalaro

Rosemary S. Vatalaro

Additionally, the authors acknowledge Yasuhiro Monden for documenting various kanban equations in use at Toyota.

INTRODUCTION

This text takes you through the step-by-step kanban implementation process by telling the story of how kanban was implemented by a fictitious golf club manufacturer—Emca Golf. Although the name of the company is fictitious, the implementation process and the issues that surface during it are not. The story of Emca is based on the authors' many years of highly successful, on-site kanban pull implementation experience in companies much the same as yours.

KANBAN PULL

The concept of kanban pull has been around for quite a while. Its origins date back to the late 1940s, when Taiichi Ohno, then the manager in charge of Toyota Motor Company's machine shop, was conducting some product replenishment experiments. Much of his time was spent dealing with the frustrations caused by ongoing product shortages. Taiichi was attempting to adapt the concept of American-style supermarket replenishment (recently transplanted into Japan) to his machine shop's processes. After much trial and error, he made the connection between just-in-time production and kanban pull.

Kanban pull is a simple, but very powerful, concept. In its purest form, it is a visual replenishment signaling system that effectively "connects" the supplying and consuming processes that exist throughout the entire value stream. Kanban pull is so effective because it is based on real-time, actual consumption, not hypothetical production forecasting or prediction schemes (which are usually wrong). If your value stream is based on, or includes, any of the following attributes, you have much to gain from implementing a kanban pull system:

- Component fabrication and supply
- Assembly and subassembly
- Manual production cells
- Semi-automated and manual lines
- Low-volume production

- High-volume production

- High and low mix

Kanban pull is one of many techniques that comprise a lean production system. As is the case with any component of a larger system, kanban has prerequisites that must be met and interdependencies that must be understood in order for it to be able to achieve and sustain optimal results.

Chapter Overview

The subject matter of this text clearly explains how you should proceed with implementing kanban pull in your manufacturing operations.

In Chapter I, "Preparation Work," we describe the prerequisites and interdependencies. We explain how to create a strong foundation upon which to build your kanban pull system. It is here that we expose the interrelationship between kanban and other lean concepts, such as value stream mapping, product family definitions, continuous flow production, set-up reduction, process/equipment reliability, and the inevitable human issues dealing with the willingness to unlearn past practices. We also address the cultural discipline required to maintain and manage continual incremental improvement, the essence of *kaizen*.

With a strong foundation built, we are well positioned to take full advantage of controlling production and replenishment using a kanban pull system.

In Chapter II, "The Principles of Kanban," we explain the basic elements and mechanics of a kanban pull system. We start with reviewing the various forms the kanban signal can take and the factors that influence which signaling approach is best for your application. We discuss the five essential rules that govern a kanban pull system's functioning. From there, we discuss the design, functionality, and different types of kanban cards and how they interact with the kanban board and supermarket subsystems. We thoroughly explore the purpose, design and mechanics of kanban boards and explain how all of the above collaborate in assuring that you produce the right product at the right time in the right quantity. Once you understand these various subsystems that form a kanban pull system, you can take the next step, which involves some "number crunching."

The focus of Chapter III, "The Numbers," is to establish the number of kanban (and thus the amount of inventory) needed in the kanban pull system envisioned on the future state value stream map. We explain in detail how to perform a requirements study and A-B-C-D classification, how to allocate production resources to specific products, how to compensate for shared supplying resources, how to determine order frequency, and how to determine and allocate the set-ups available. We finish Chapter III by exploring the multiple facets of the kanban equation and explaining the mechanics of how to apply all of the previously collected data to the equation to determine the number of kanban needed for each product to be placed under kanban control. After having analyzed all the facts and data, it is time to "Go-Do" or act upon the analyses to implement kanban pull.

In Chapter IV, "Some Tools to Help You 'Go-Do,'" we outline a recommended kanban pull implementation strategy. We explain the logic of first focusing internally in developing a competency with implementing kanban pull systems before extending it to your suppliers and customers. Also, in this chapter we provide you with "the kaizen kit." The kit has everything you need to conduct a highly successful kanban kaizen event. We include detailed agendas and planning check sheets, event facilitation guidelines, and blank forms.

The accompanying CD includes many of these forms, as well as the calculation worksheets used throughout the book.

Once you have implemented your kanban system, you will need to make sure you have a plan to manage it on an ongoing basis. In Chapter V, "Managing the System," we cover what you should consider when establishing a plan to manage the daily operation of the system. This includes factors for managing both normal and abnormal conditions.

Keeping in mind the old adage that "forewarned is forearmed," in Chapter VI, "Frequently Asked Questions," we explain some issues of your kanban pull implementation that you might not anticipate. These include the need to prepare an inventory "burn-down" strategy and the way to deal with typical performance measures conflicts that invariably arise. Additionally, we outline what should be included in your kanban pull procedure. It is here that you will document how your new kanban-based production control system will operate and how it will be managed. Also in this chapter we prepare you in advance for the issues that you will run into concerning how your existing Information Technology should (and should not) be integrated with your kanban pull system.

GETTING STARTED

Before we get started, a few very important points to keep in mind:

First, the implementation of kanban is not easy and there are no short cuts. Most of the difficulty will come from changing the paradigms of the people in your organization. That is the bad news. The good news is, the results of implementation are well worth the effort. In fact, we consistently find that once your new kanban pull system is up and running, your people will never go back to past practices.

Secondly, because kanban is rarely the first lean tool applied in the lean transformation process, we make the assumption that the reader has a basic understanding of the concepts of waste (muda) elimination, value stream mapping, and continuous flow production. A good understanding of these three concepts will facilitate the understanding and use of this text. Please refer to the Recommended Reading section in the back of the book for additional information.

Lastly, an assumption made in this text is the user's ability to successfully deploy cross-functional teams in the kanban pull implementation or "Go-Do" phase. Cross-functional teams lead the most successful lean implementations. As is the case with any lean tool implementation, no one person in your organization possesses all of the needed knowledge or perspectives. The involvement of the users of your new kanban pull system (hourly operators, material handlers, etc.) is mandatory, as they are the people that will make the system work on a day-to-day basis. The kanban pull system will require materials and factory management to closely collaborate with production associates on the following: system design, launch, management, trouble-shooting, and future improvement activity.

Although just a few people in your organization will need to understand kanban pull fully to lead the implementation effort and train others, everyone in the company needs to be trained on the basic functionality of your new kanban pull system.

Key team members who will need to roll up their sleeves and physically implement kanban on the factory floor will include people from:

- Production
- Scheduling
- Material management and handling
- Production control
- Industrial and manufacturing engineering
- Purchasing
- Suppliers
- Customers (as appropriate)

CHAPTER 1

Preparation Work

❏ What does the current state value stream map look like?

❏ What is the takt time?

❏ What could the future state value stream map look like?

  • Where should the product flow continuously?

  • Where should the product be "pulled" via kanban?

To begin, here is an introduction to the discrete products manufacturer case used to guide you through the process of implementing a mixed model kanban pull system.

The Company

Emca Golf Club Company
"The Sticks That Champions Use"

A Division of Emca Sporting Goods

Emca Sporting Goods is a manufacturer headquartered in Greeneville, USA. They have four strategic business units: Golf, Water Sports, Racket Sports, and Camping Equipment. Emca's top management recognizes the competitive advantages associated with improved product velocity through waste elimination. As such, several months ago, Corporate determined their initial lean implementation would be in the high-end golf club business unit. This business unit is the largest at Emca, as it accounts for 45 percent of corporate revenues.

Business Units	Golf	Water Sports	Racket Sports	Camping Equipment
% Annual Corp. Revenue	45%	25%	20%	10%
Annual Corp. Revenue	$45,000,000	$25,000,000	$20,000,000	$10,000,000

Emca Golf is recognized as the world leader in innovative golf club design. However, they have some particularly threatening problems too. For several years, they have been plagued with very high inventory levels, not much working capital, and very long manufacturing lead times. To make matters worse, their assembly operations have been inundated with frequent component shortages. *Emca has plenty of the parts they do not need and few of the parts they do need.* No matter what they do, they are not able to make the right parts at the right time. Attempts at better controlling production using sophisticated computer hardware and software have proved ineffective (even considered countereffective by some). The result has been many meetings to discuss "what are we going to make today?"

and frequent arguments among expediters. The problems continue while management and the workforce become more disheartened with the passing of each day. The result is poor delivery performance and a slipping market share trend.

MARKET SHARE HISTORY (5-Year Trend)					
FY Ending	2002	2003	2004	2005	2006*
Market Share	57%	46%	41%	37%	34%

* = Projection

EMCA GOLF'S PRODUCTS AND PROCESSES

Emca Golf is a manufacturing division that produces four different lines of golf clubs. They are: the "Sinker" putter, the "Blaster" pitching wedge, the "Eagle" irons, and the "Redhawk" line of drivers.

The equipment and processes used to produce their club components are highly specialized. They are all designed specifically to machine golf club heads, mold grips and fabricate shafts, and assemble the final product. Emca currently uses two computer numerical control (CNC) machining centers for head production.

WHAT DOES EMCA'S CURRENT STATE VALUE STREAM MAP LOOK LIKE?

Emca Golf acknowledges the need to improve, but needs to understand where to start the improvement process. To help with these improvements they sent their "lean champion" to a value stream mapping seminar to learn the techniques. Immediately upon his return, he developed a current state value stream map (page 4). A quick analysis of their current state map indicates a production approach based on push, process islands, batch and queue, high inventory, and no discernable flow at all. The results are very long lead times and a general unresponsiveness to changes in customer demand. (The value stream mapping icon definitions can be seen on page 95.)

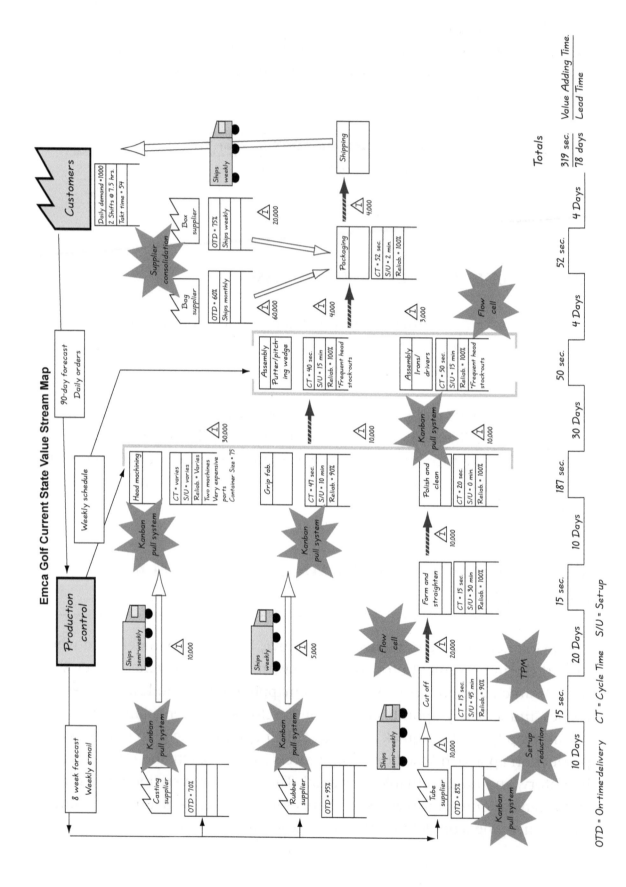

Emca Golf Current State Value Stream Map

VALUE STREAM MAPPING GUIDELINES

The following information is the minimum needed to help establish the area of initial improvement focus. Be certain to take the necessary time to capture accurate data, as the quality of your decisions is a function of the quality of the information you base them upon. All too often numbers are accepted based on "tribal knowledge" that turns out to be inaccurate. Later on, as part of the *kanban kaizen* event planning process (reviewed in Chapter IV), additional detailed information will need to be collected:

- **Machine cycle times.** These will be needed in order to perform a requirements study and determine order frequency, as outlined in Chapter III.

- **Set-up times.** The current state set-up times will be needed in order to determine how frequently we can set up our processes.

- **Process reliability.** This information will be needed to establish the actual capacity of a supplying process. Mean Time Between Failure (MTBF), Mean Time To Repair (MTTR) and Overall Equipment Effectiveness (OEE) are the metrics most commonly deployed here.

- **Number of machines and operators.** Required for all processes in the value stream.

- **Container sizes.** These will be needed for the kanban calculation process.

- **Available production time.** This is the basis upon which we determine capacity in the takt time calculation.

- **Customer demand levels.** Expressed as the current average daily demand for each item.

- **Defect rates.** Also needed to support order frequency determination.

- **Work in process levels (quantities and dollars).** These are needed in order to estimate the current manufacturing lead time, and quantify future state cash flow improvement.

- **Chronic shortage resources.** This data will help identify pilot product candidates for kanban control.

- **Communications.** Illustrate the current production control communication methods and frequency.

- **Takt time.** A very important component, which is explained in more detail below.

WHAT IS THE TAKT TIME?

Takt means beat or rhythm. In the context of production, takt time is the heartbeat of production. It is a calculated number that indicates the pace of customer consumption. The equation is expressed as:

Available Production Time / Customer Demand

Takt time is usually expressed in seconds and expresses the frequency of customer consumption.

Applying the takt time equation for Emca Golf's combined product lines, the result is:

54000 daily available seconds / 1000 units of daily demand =
54-second takt time (or 1 unit consumed every 54 seconds)

Remember that production operations designed to be equal to or less than takt time are reliable, predictable, and stable—all attributes that facilitate the implementation of a kanban pull system.

WHAT COULD THE FUTURE STATE VALUE STREAM MAP LOOK LIKE?

Emca's lean champion organized and delivered a presentation to management to inform them of the current state findings. They were shocked to see the 78-day manufacturing lead time and frequent stock-outs at assembly caused by head fabrication. The lean champion then described the opportunity for compression of the manufacturing lead time and inventory reduction. During the presentation, one of the managers stated, "Yes, it looks bad in theory, but our business is different. There are reasons why we do business the way we do." The lean champion found it convenient to quote his value stream mapping trainer—"If you always do what you have always done, you will always get what you have always gotten." The lean champion, armed with the value stream facts, discovered that the change-resistant forces within Emca Golf could mount little resistance.

It was obvious what had to happen next. The lean champion collaborated with the management team and developed a future state map for their golf club lines (page 8). This future state map reflects what Emca Golf wants to become. They understood it would most likely be a tiered or iterative process to achieve their goal. By allowing the value stream map to guide them, they knew where they had to begin.

WHERE SHOULD THE PRODUCT FLOW CONTINUOUSLY?

Reviewing the future state map, several opportunities presented themselves. One major discovery was the ability to create one-piece flow in the shaft fabrication portion of the value stream. Another big discovery was the identification of where one-piece flow was simply not possible. For example, the head machining process must supply 16 different head models to final assembly. The table below lists the 16 golf club head models.

HEAD PART NUMBER ANALYSIS				
Club Type:	Putter	Pitching Wedge	Irons	Drivers
Head Part Number:	PL10	PWL10	IR10	DR10
	PR20	PWR20	IR20	DR20
			IR30	DR30
			IL10	DL10
			IL20	DL20
			IL30	DL30

Note: An "R" in the part number designates a right-handed club head and an "L" designates a left-handed club head.

Thinking back to his value stream mapping training, the lean champion remembered that when flow has to be broken, the next best approach in controlling production is to pull by way of a kanban pull system. This new kanban pull system would replace the daily production control functionality of Emca Golf's computer scheduling system. This is the system that has proven to be so expensive (financially and emotionally) and ultimately ineffective, primarily due to data integrity related issues.

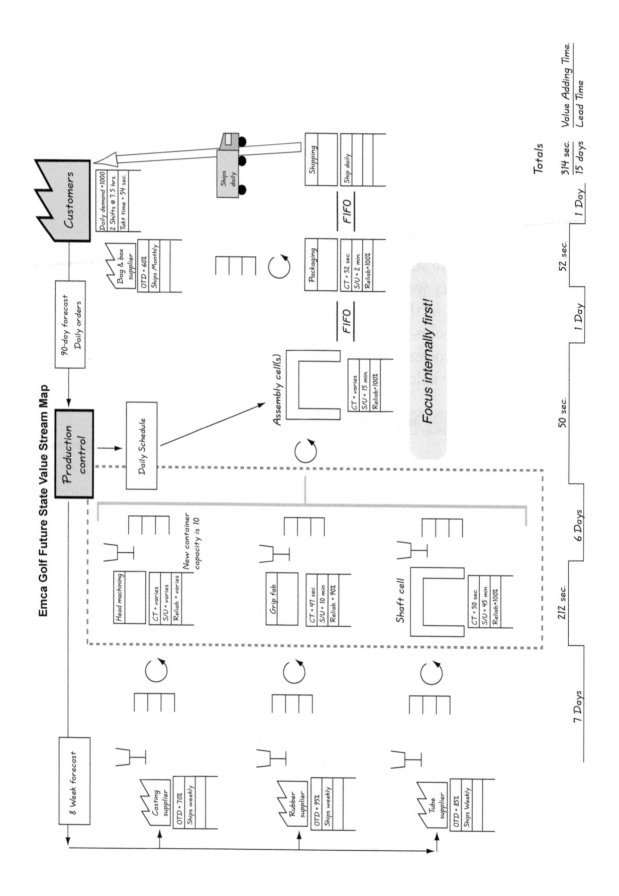

Emca Golf Future State Value Stream Map

Demonstrating a bias for action and having learned the process of *kaizen*, the lean champion planned for, organized, and conducted a series of kaizen events. He started with 5S kaizen events, which helped instill a sense of change acceptance and discipline in the workplace. He then refocused the kaizen team on the shaft fabrication process. Emca Golf's kaizen team deployed the tool of cell design in the shaft fabrication process and created "one-shaft flow."

The team then noticed that shaft production became much more stable and predictable.

SHAFT METRICS COMPARISON SUMMARY		
	Pre-Kaizen	Post-Kaizen
Part travel distance	3,500 ft.	255 ft.
Space (sq. ft.)	45,000	17,000
# units of inventory	30,000	1,000
$ value of inventory	$1,094,100	$36,740
Lead time	30 days	1 day

WHERE SHOULD THE PRODUCT BE "PULLED" VIA KANBAN?

Emca Golf decided to first focus internally with implementing kanban. They recognize they must first achieve an internal competency with kanban before they can extend it to their suppliers and customers.

Emca Golf is now well positioned to begin implementing a kanban pull system. Due to the large disruption at club assembly caused by golf club head shortages, the location to implement kanban pull first is between the head machining and final assembly processes. Here again Emca will deploy the principles of kaizen and organize a kanban kaizen event team. The first challenge of the team will be to gain a better understanding of the principles of kanban.

CHAPTER II
The Principles of Kanban

❏ What are the fundamental components and mechanics of a mixed model kanban pull system?

The Emca Golf lean champion gathered the members of the kanban kaizen team and reviewed the following information so that all members understood exactly how kanban works.

KANBAN PULL SYSTEMS TECHNOLOGY

The literal translation of the Japanese word kanban is "signal" or "sign board." In the context of production, it refers to a visual production control system that signals replenishment. The replenishment signal itself can take a variety of forms, from cards to ping-pong balls to rings on a pegboard. What makes kanban so powerful is the fact that this signal is generated by actual consumption, what we refer to as the pull of the downstream customer. Kanban connects processes together in the value stream by making the pull of the customer (or consuming process) visual to the upstream supplying resource. To visualize a kanban pull system, imagine a series of connected loops, like a chain. By the customer pulling on one loop at the end of the chain, all the other loops upstream "feel" the same pulling action.

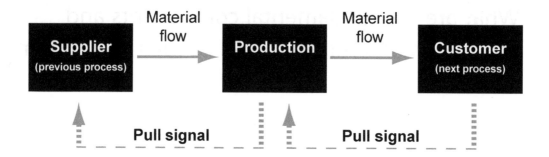

Virtually any product can be the subject of this pulling: hardware, parts, subassemblies, and final assemblies. A properly designed and implemented pull system results in the right parts being produced at the right time in the right quantity throughout the entire value stream.

Several factors influence the decision of what form the pull signal should take. Key considerations include:

- The distance the signal has to travel

- Speed (what is the effect of signal velocity?)

- Signal content (what information must be communicated?)

As stated previously, Emca decided to first implement kanban at the head machining-to-assembly cell loop, as illustrated on their future value stream map (page 8).

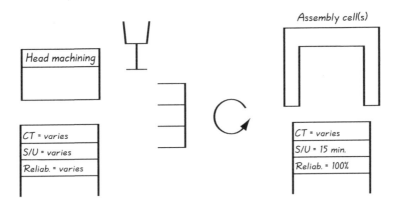

This means the production "motivation" at the head machining center will change. In the past, the center produced parts when a scheduling report or an expediter authorized it to do so. Now, the center will commence production only when a consumption signal is received from the consuming process or customer (even if the material, an operator, and a machine are available).

An important point to remember is that a kanban pull system controls the activity of the *supplying resource*. As such, the supplying resource receives the kanban pull signal to replenish a product it is responsible for producing. A great degree of freedom exists in defining a resource. A supplying resource could be defined as a single machine, a group of identical or similar machines, a fully integrated and self-contained cell, or an external vendor. The key point is that the supplying resource receives the kanban pull signal indicating that consumption occurred. A complete product can be produced in any of the ways mentioned previously.

For Emca, the team thinks it makes sense to establish the two CNC machines used for head production as two separate resources. The team thinks each will be dedicated to the machining of specific golf club heads.

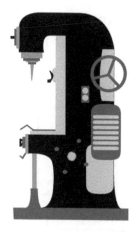
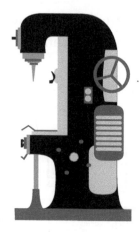

Head CNC Machining Center 1   Head CNC Machining Center 2

Resources whose production activity is controlled by kanban pull must recognize and adhere to some significant new rules of production. They are:

- The consuming process consumes only what is needed.

- The supplying process produces only what the kanban pull system authorizes, when it authorizes it. Emca Golf could have a machine available, an operator available, the time and the raw material available, but without a signal from the kanban pull system, NO PRODUCTION IS AUTHORIZED!

- Defective products will not be sent to the consuming process from the supplying process.

- Kanban should reflect changes in demand, as demand is frequently subject to trends.

- The number of kanban should be minimized over time as they buffer waste in your value stream.

- These rules are not negotiable.

KANBAN CARDS

The most frequently used signaling method is the kanban card because it is an adaptable and flexible method. There are two different types of kanban cards. They are the *permanent kanban card* and the *single use kanban card*.

Permanent Kanban Cards

Permanent kanban cards are cards that will be continuously recycled through the kanban pull system. Inventory for the products authorized by these cards will usually reside in the "supermarket" (discussed later in this chapter). These products will be the items the customer typically consumes at predictable time intervals.

In designing the permanent kanban cards for the new kanban pull system, Emca recognized the need to include the following minimum information:

- Material, part, subassembly, or assembly identifier
- The internal or external supplying process
- The consuming processes
- The container quantity
- The supermarket "address"
- A card serial number (facilitates auditing cards to identify missing cards)
- An illustration of the part

NOTE

Avoid serializing cards with a "# of #" scheme, i.e., 1 of 8, 2 of 8, etc. If circumstances warrant changing the number of cards in the system, ALL of the cards will need to be replaced—a very wasteful procedure. Instead, just indicate card 1, card 2, card 3, etc., and keep a separate record of how many cards exist for that item.

To help facilitate "visual control," Emca Golf's kanban cards will make use of color-coding at multiple levels.

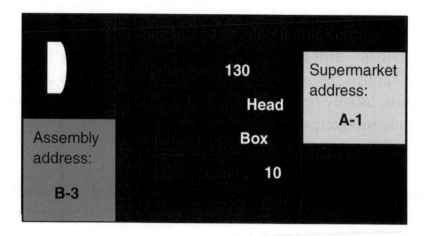

To visually manage the new kanban pull system, Emca Golf's kanban kaizen team decided that the kanban cards for the head machining supplying resource would be red (the "base" color). The area at right designates where the item is located in the supermarket. The area at left indicates the consuming process is assembly.

Keeping the inventory information with the inventory is a key feature of a kanban pull system. As such, kanban cards will be attached to standardized containers, such as the following:

- Pallets

- Bins

- Trays

- Boxes

- The actual workpiece itself

In the case of Emca, the team decided new, right-sized plastic containers (color-keyed to the kanban card base color) would be used to hold the machined club heads. *Emca Golf's future state value stream map indicates that a new head container quantity of 10 units may work well.* The kanban card will be placed into a clear plastic sleeve firmly attached to each container.

> **NOTE**
>
> *In some instances, it might make sense to have a single kanban card represent multiple containers—for example, having multiple containers on a pallet with one kanban card for the entire pallet. There are two important issues that factor into the decision. First, you must not lose the ability to determine if the inventory in the value stream is authorized once the first container is opened and the kanban card has been removed. Second, be aware of the impact your decision will have on the number of kanban in the value stream. Too many or too few kanban can make the system unmanageable.*

Another consideration for the kanban kaizen team concerns the rules as to when the kanban card must be removed from the container. Typically, three options are available:

- **Full container rule**—as soon as the first product is consumed from the container, the kanban card is removed and returned to the kanban board.

- **Mid-container rule**—as soon as half the products in the container are consumed, the kanban card is removed and returned to the kanban board.

- **Bottom of container rule**—as soon as the container is emptied, the kanban card is removed and returned to the kanban board (*see* page 21).

The Emca team has initially decided to use the bottom of container rule as they feel it lends itself better to visually controlling the inventory. A container with any inventory in it will always have a kanban card with it. Accordingly, any container found containing inventory that is missing a kanban card will be suspected of being "unauthorized."

The team recognizes that using the bottom of container rule is a trade-off. This is because the sooner the consuming process removes the kanban card and sends it back to the supplying process, the sooner the supplying process will be replenished. This will facilitate keeping the value stream inventory levels as low as possible and provide the fastest replenishment lead time.

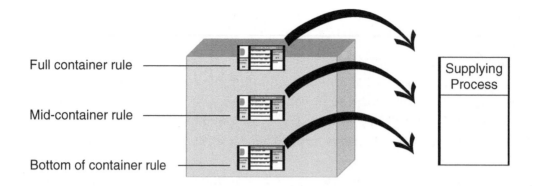

Full container rule

Mid-container rule

Bottom of container rule

Supplying Process

The Single Use Kanban Card

Single use kanban cards offer a way to deal with products where the customer demand does not warrant maintaining the inventory that permanent cards would authorize. With single use kanban cards, the card is introduced into the kanban pull system only after a firm customer order is placed. Replenishment is prevented as the kanban card is extracted from the kanban pull system as soon as consumption occurs. Circumstances that frequently warrant the use of single use kanban cards include:

- When a "spike" increase in demand occurs.

- When producing products that are very infrequently ordered.

- When producing special order items.

- For excess inventory burn-down (frequently resulting from implementing kanban pull).

Single use cards are often specially color-coded so they stand out. With the exception of the supermarket location, all of the information specified on a regular kanban card is also specified on a single use kanban card.

SUPERMARKETS AT EMCA GOLF

Supermarkets help tie together the producing and consuming processes in a kanban pull system by buffering the demand signals coming back from the consuming processes. They contain inventory that was authorized by a pull signal from the internal or external customer. Without supermarkets, every signal coming back to the supplying process would mean immediate replenishment is required. In a high-mix manufacturing environment, the supplying process most likely would not be able to keep up. Supermarkets are very different from queues. Emca Golf's traditional queues or accumulations of inventory were the result of a computer system trying to push product through the value stream. In a kanban pull system, the supermarket will be one of three places inventory will be authorized to reside. The other two are point of use at the consuming process or in transit from the supplying process.

Emca Golf's consuming process, golf club assembly, will withdraw machined heads from the supermarket, and the head machining work center will replenish the supermarket when signaled to do so by the kanban board.

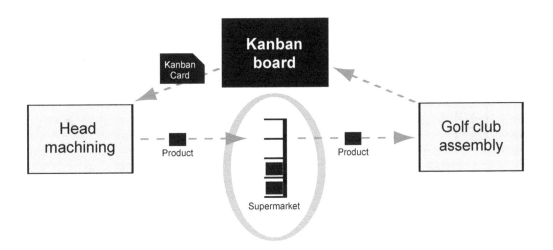

Emca Golf's future state value stream map makes extensive use of supermarkets in their new kanban pull system. The chief focus will be at head machining. The quantity of inventory residing in the supermarket will be the average expected inventory level minus the inventory positioned at the point of use at the consuming process. In the following chapter we will review the process for determining the number of kanban needed for each product placed under kanban control.

The kanban kaizen team needs to make the decision about where to physically position the supermarket in their plant. Three options exist. They are:

- At head machining.

- At assembly.

- At a point in between the two.

The best choice is the position that offers the least waste. Key factors the team will need to consider in determining the amount of waste include:

- The number of supplying workstations versus the number of consuming work stations.

- Travel distance.

- Space availability.

- The transport mechanism currently available.

- If the supplier is internal or external.

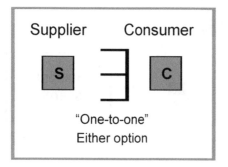

"One-to-one"
Either option

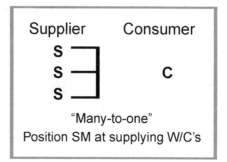

"Many-to-one"
Position SM at supplying W/C's

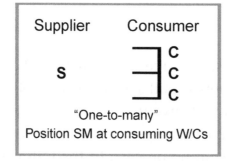

"One-to-many"
Position SM at consuming W/Cs

SM = Supermarket
W/C = Workcenter

KANBAN BOARD

A kanban board is a visual display. It is here that the kanban cards accumulate after they have been removed from empty product containers. The accumulation of cards continues until a predetermined number of cards is achieved. This quantity is displayed by the positioning of the order frequency card (*see* page 23). Emca has developed a very robust, flexible, and informative format for their kanban board that will be the foundation for all their boards. It is called the "common authorization line" design because the movable order frequency cards are used to establish the common authorization line (*see* page 22).

Although we recommend the common authorization line design, kanban boards come in infinite varieties. A robust kanban board design is adaptable, scaleable, and visually communicates, at a minimum, the following key information:

- The locations to position kanban cards in columns or rows, by product, as they are returned from the consuming process;

- A visual indicator of when replenishment is authorized and necessary, known as the authorization point;

- A visual indicator of the part number being processed, as well as the next part number to be set up and processed, "FIFO tracking";

- An area to indicate overdue status;

- An area to make notations.

Common Authorization Line

The common authorization line kanban board design functions as follows:

1. Product is consumed from containers at the consuming process. When the consuming process empties a container (remember, the team decided to use the bottom of container rule, *see* page 17), the card and container are removed from the consuming process area. The card is removed from the container. The empty container is then staged in close proximity to the kanban board.

2. The kanban card is placed on the kanban board at the next lowest open location above the order frequency card for the given part number. This process repeats until enough kanban cards representing a specific product accumulate to reach the authorization line on the kanban board.

3. Immediately upon the authorization line being reached, all kanban cards for that part number are removed and individually placed in transparent sleeves attached to empty containers. The empty containers with the attached kanban cards are delivered to the supplying process. This action authorizes replenishment by the supplying process.

4. Once the containers are filled again with completed product, the containers with the attached card are returned to their specific supermarket location (which is indicated on the kanban card). As the line stock at the consuming process is depleted, product is removed from the supermarket and delivered to the consuming process. The process repeats.

> **NOTE**
>
> *The various products produced at the supplying process must be produced in a first in/first out "FIFO" manner.*

When the supplying process has problems that delay its ability to replenish within the determined lead time, the kanban cards may accumulate past the authorization line on the kanban board. This means that the consuming process may need to consume lead time and possibly safety time inventory in order to satisfy consuming process demand. The kanban board makes this situation very "visual." If replenishment does not occur within the lead time and safety time established, a product shortage will occur at the consuming process.

Order Frequency Card

The order frequency card establishes a common authorization line for each part controlled by the kanban board. Without the order frequency card, all kanban cards would accumulate from the lowest point on the board and most likely have different authorization lines plotted on the board for them. This approach does not lend itself well to the lean concept of visual control.

NOTE

Refer to the PowerPoint file on the CD named "Animated Kanban Board" to see an animated example of how a kanban board functions.

Capacity, changeover frequency, and replenishment lead time can be manipulated by changing the position of the order frequency card. By moving the order frequency card up, the authorization line is reached faster—thus, capacity will be decreased because there will be more frequent set-ups. However, lead times will be reduced because the lot size to be manufactured will be reduced. Moving the order frequency card down will have the opposite effect of increasing capacity and increasing replenishment lead time. In the next chapter, we will review the process for calculating the order frequency card position and detail the process of determining the number of kanban needed to properly "connect" processes.

CHAPTER III
The Numbers

❒ What is a Requirements Study?

❒ What is Order Frequency Determination?

❒ What is the number of kanban needed?

❒ What is a kanban flow diagram?

NOTE

Your company culture will affect the assumptions you will make while calculating the number of kanban needed (and subsequently, your inventory levels). Decisions lie ahead for answers to questions such as:

- *"Do we (remember, this is a team effort) round up or round down or don't round at all?" and "How much safety inventory (safety time) do we really need?"*

The most important question to consider is:

- *"How aggressively do we want to drive waste from our operations?"*

WHAT IS A REQUIREMENTS STUDY?

Once Emca Golf's kanban kaizen team understands the basic principles of kanban and has completed all the necessary kanban pull systems preparation work we discussed in the previous sections, they can begin the process of determining the number of kanban needed to effectively link head machining and assembly.

To start out, the team must identify *all* of the golf club heads that are produced in head machining. The kaizen team identified the following part numbers:

HEAD PART NUMBER ANALYSIS				
Club Type:	**Putter**	**Pitching Wedge**	**Irons**	**Drivers**
Head Part Number:	PL10	PWL10	IR10	DR10
	PR20	PWR20	IR20	DR20
			IR30	DR30
			IL10	DL10
			IL20	DL20
			IL30	DL30

We can see from the above table that head machining is a shared resource that produces a variety of 16 different heads for club assembly.

NOTE

Kanban is very appropriate in controlling shared supplying resources.

Having identified the 16 part numbers that will be placed under kanban control, the kaizen team will need to quantify the amount of daily demand each head model places on head machining in order to satisfy the needs of its internal customer—assembly. This is called performing a *requirements study* and the tool we use is called the Requirements Study Worksheet.

				REQUIREMENTS STUDY WORKSHEET			
			COMPONENT: HEADS				
Component Part Number	Average Daily Demand	A-B-C-D Designation	Machine Center 1 Cycle Time	Machine Center 2 Cycle Time	Selected Machine Center	Machine Center 1 Loading	Machine Center 2 Loading
					TOTALS:		

The first step for completing the Requirements Study Worksheet is filling in the column labeled Component Part Number. The next step is to determine a current average daily demand for each head by analyzing historical order patterns as well as current booked orders. The team did this, and, as a result, the Requirements Study Worksheet looks as follows:

			REQUIREMENTS STUDY WORKSHEET				
			COMPONENT: HEADS				
Component Part Number	Average Daily Demand	A-B-C-D Designation	Machine Center 1 Cycle Time	Machine Center 2 Cycle Time	Selected Machine Center	Machine Center 1 Loading	Machine Center 2 Loading
PL10	25						
PR20	135						
PWL10	29						
PWR20	111						
IR10	100						
IR20	100						
IR30	100						
IL10	75						
IL20	75						
IL30	50						
DR10	60						
DR20	20						
DR30	40						
DL10	30						
DL20	30						
DL30	20						
					TOTALS:		

The next step for the kaizen team is to categorize each head product by its average daily demand level. We shall use a method referred to as A-B-C-D Analysis to accomplish this.

A-B-C-D Analysis

The Requirements Study Worksheet is where the team characterizes the level of demand for each golf club head model. Relative to one another, golf club heads are either "A" items, "B" items, or "C" items. "A" items are frequently ordered, high-volume items. "B" items are ordered somewhat frequently and are medium volume. "C" items are infrequently ordered or low-volume items.

NOTE

In the case of special order items, (at Emca Golf, the head machining resource produces none) you would classify them as "D" items and assign single use kanban cards for their production.

The Emca kanban implementation team agreed upon the following parameters in establishing their A-B-C-D boundaries for the heads:

Classification	Daily Demand
A	≥ 76
B	≤ 75 and ≥ 51
C	≤ 50

Applying the above parameters, the team designated the various club heads as follows.

REQUIREMENTS STUDY WORKSHEET							
COMPONENT: HEADS							
Component Part Number	Average Daily Demand	A-B-C-D Designation	Machine Center 1 Cycle Time	Machine Center 2 Cycle Time	Selected Machine Center	Machine Center 1 Loading	Machine Center 2 Loading
PL10	25	C					
PR20	135	A					
PWL10	29	C					
PWR20	111	A					
IR10	100	A					
IR20	100	A					
IR30	100	A					
IL10	75	B					
IL20	75	B					
IL30	50	C					
DR10	60	B					
DR20	20	C					
DR30	40	C					
DL10	30	C					
DL20	30	C					
DL30	20	C					
					TOTALS:		

Resource Allocation

We established previously that the head machining resource consists of two similar, but not identical, machine centers. During the resource allocation process, the team will determine which heads will be produced on which machine. The Emca team will be guided by their A-B-C-D analysis findings.

The machine cycle times for each golf club head were determined and are shown below.

MACHINE CYCLE TIME ANALYSIS		
Head Part Number	Machine Center 1 Cycle Time (sec.)	Machine Center 2 Cycle Time (sec.)
PL10	45	36
PR20	45	36
PWL10	50	40
PWR20	50	40
IR10	65	30
IR20	65	30
IR30	65	30
IL10	65	30
IL20	65	30
IL30	65	30
DR10	120	85
DR20	120	85
DR30	120	85
DL10	120	85
DL20	120	85
DL30	120	85

A quick analysis indicates that regardless of which product is being produced, Machine Center 2 has a faster cycle time than Machine Center 1. As such, Machine Center 2 is the machine that Emca will want to run most, if not all, of their "A" and "B" items. These higher volume items are typically the leverage points in being able to maximize product velocity. In doing so, Emca will maximize inventory turnover performance and minimize manufacturing lead time for these items. Accordingly, Machine Center 1 has a slower cycle time, but also provides for a shorter set-up time. The team rightly determined Machine Center 1 should be used for the "C" items. The Machine Allocation Table, shown below, summarizes their initial decisions about which product should be run on which machine.

MACHINE ALLOCATION TABLE			
Head Part Number	**A-B-C-D**	**Machine Center 1**	**Machine Center 2**
PL10	C	X	
PR20	A		X
PWL10	C	X	
PWR20	A		X
IR10	A		X
IR20	A		X
IR30	A		X
IL10	B		X
IL20	B		X
IL30	C	X	
DR10	B		X
DR20	C	X	
DR30	C	X	
DL10	C	X	
DL20	C	X	
DL30	C	X	

Using the data the team has collected up to this point, they can complete the remainder of the Requirements Study Worksheet (on the next page) and determine if they have overloaded either of their two available machining centers.

NOTE: TOTAL DEMAND

In designing a kanban pull system, you must understand, and factor into the requirements study, the total demand placed on the supplying resource. Products that are outside of the initial implementation focus may be produced at the supplying resource. For the near term, these products will be controlled using the traditional methodology. From a daily operating perspective, this will result in using dual production control systems until all products are under kanban control. This dual replenishment approach is to be expected in the early stages of kanban implementation. The unadvisable alternative is to "swallow the elephant whole," taking on more change than you can handle at one time. Attempting a plantwide kanban implementation all at once may jeopardize a powerful improvement technology. Our advice—do not take on too much all at once.

REQUIREMENTS STUDY WORKSHEET							
COMPONENT: HEADS							
Head Part Number	Average Daily Demand	A-B-C-D Designation	Machine Center 1 Cycle Time (sec.)	Machine Center 2 Cycle Time (sec.)	Selected Machine Center	Machine Center 1 Loading*	Machine Center 2 Loading*
PL10	25	C	45	36	1	1125	
PR20	135	A	45	36	2		4860
PWL10	29	C	50	40	1	1450	
PWR20	111	A	50	40	2		4440
IR10	100	A	65	30	2		3000
IR20	100	A	65	30	2		3000
IR30	100	A	65	30	2		3000
IL10	75	B	65	30	2		2250
IL20	75	B	65	30	2		2250
IL30	50	C	65	30	1	3250	
DR10	60	B	120	85	2		5100
DR20	20	C	120	85	1	2400	
DR30	40	C	120	85	1	4800	
DL10	30	C	120	85	1	3600	
DL20	30	C	120	85	1	3600	
DL30	20	C	120	85	1	2400	
					TOTALS:	22625	27900

Key Point: Machine selection was based on the faster machine, Machine Center 2, running the higher volume ("A"/"B") items.

* Machine Center loading calculation: Daily demand $\times$ Machine cycle time = Machine loading. This is calculated for each part and then totaled by machine. For example, for PL10, multiply the daily demand of 25 by the cycle time for Machine Center 1, which is 45 seconds. Therefore, machine loading = 25 $\times$ 45 = 1125 seconds.

Emca is now ready to move on to the next phase of the kanban implementation process, "Order Frequency Determination."

WHAT IS ORDER FREQUENCY DETERMINATION?

The Emca kanban implementation team needs to refer back to the value stream map and supporting data to extract the information needed to determine the actual capacity of the head machining resources. This information includes:

- The number of machines available for production,
- The number of shifts they operate on,
- The calculation of gross machine capacity,
- Machine reliability,

- A calculation of net machine capacity,

- Cycle time data,

- Set-up data.

The following head machining equipment analysis summarizes the above information for the machining centers.

HEAD MACHINING EQUIPMENT ANALYSIS							
	Capacity per Shift (sec.)	Number of Shifts	Gross Machine Capacity (sec.)	Actual Capacity	Net Machining Capacity (sec.)	Set-Up Time (min.)	Set-Up Time (sec.)
Machine Center 1	27000	2	54000	60%	32400	15	900
Machine Center 2	27000	2	54000	70%	37800	30	1800
Total Net Machining Capacity:					**70200**		

The team determined that Emca has 54,000 seconds of gross available capacity at each machine center. As the total demand for both machine centers on the Requirements Study Worksheet (50,525) did not exceed 108,000 (54,000 × 2 machines) seconds, Emca has sufficient gross capacity. However, the team needs to look at this more carefully and understand the *net* (true) capacity and the impact of waste in the machining processes. The team will accomplish this by performing a process called Order Frequency Determination.

Order frequency is the frequency at which Emca plans to set up and order each golf club head. Determining the order frequency is a four-step process.

1. The first step in the process of determining the order frequency is to calculate the time available per day for Emca to set up each head machine center. To do this the team will need to refer back to the data captured on the Head Machining Equipment Analysis table and Requirements Study Worksheet. From the Head Machining Equipment Analysis table "Net Machining Capacity" number we subtract the "Machine Loading Time" number provided on the Requirements Study Worksheet. The following table illustrates the calculation process and result for each machine.

SET-UP OPPORTUNITIES CALCULATIONS

	Net Machining Capacity (sec.)	— Machine Loading Time (sec.)	= Time Available to Set Up (sec.)
Machine Center 1	32400	22625	9775
Machine Center 2	37800	27900	9900

2. The second step in the process is to divide the Time Available to Set Up by the set-up time recorded on the Head Machining Equipment Analysis table. This will yield the number of set-up opportunities per day Emca has at each machine.

	Time Available to Set Up (sec.)	/ Set-up Time (sec.)	= Number of Set-up Opportunities/Day
Machine Center 1	9775	900	10.8
Machine Center 2	9900	1800	5.5

3. The third step for the Emca team is to distribute the available set-ups to the products we determined in the requirements study would be made by each machine (*see* the Machine Allocation Table, shown here again.)

MACHINE ALLOCATION TABLE			
Head Part Number	A-B-C-D	Machine Center 1	Machine Center 2
PL10	C	X	
PR20	A		X
PWL10	C	X	
PWR20	A		X
IR10	A		X
IR20	A		X
IR30	A		X
IL10	B		X
IL20	B		X
IL30	C	X	
DR10	B		X
DR20	C	X	
DR30	C	X	
DL10	C	X	
DL20	C	X	
DL30	C	X	

There is no equation to apply for this process. There are, however, three guiding considerations for the Emca team to apply.

Guiding Principle # 1: Distribute the set-ups such that they are roughly proportional to the demand level of the item. This means that the heads classified as "A's" will be set up and ordered more frequently than the heads classified as "B's." The same applies for "B" heads relative to "C" heads.

Guiding Principle # 2: Leave some set-ups undistributed so that the time value they represent can be used for the production of "Special Order" requirements. This is called "discretionary capacity."

Guiding Principle # 3: It is perfectly acceptable to distribute fractions of a set-up. For example, distributing 0.5 set-ups per day to an item means we intend to set up and run the item every other day.

By applying these principles to the products produced on the head machining centers, the Emca team determined the following set-up distribution:

SET-UP DISTRIBUTION TABLE					
MACHINE CENTER 1			**MACHINE CENTER 2**		
Head Model #	*Average Daily Demand*	*Distribution*	*Head Model #*	*Average Daily Demand*	*Distribution*
IL30	50	2	PR20	135	1
DR30	40	1	PWR20	111	0.75
DL10	30	1	IR10	100	0.5
DL20	30	1	IR20	100	0.5
PWL10	29	0.5	IR30	100	0.5
PL10	25	0.5	IL10	75	0.25
DR20	20	0.5	IL20	75	0.25
DL30	20	0.5	DR10	60	0.25
Total Allocations		7.0	Total Allocations		4.0
Total Opportunities		10.8	Total Opportunities		5.5
Discretionary		3.8	Discretionary		1.5

Using these distributions leaves Emca with a discretionary capacity of 3.8 set-ups (equaling 3420 seconds for 10.56% of net machine capacity) at Machine Center 1 and 1.5 set-ups (equaling 2700 seconds for 7.14% of net machine capacity) at Machine Center 2.

4. The fourth and final step in the process is to divide each number in the Distribution No. column into the number one. The resultant number is the "Order Frequency" of the golf club head model. The tables below show the results of this calculation process.

ORDER FREQUENCY DETERMINATION WORKSHEET					
Machine Center 1			**Machine Center 2**		
		1 / Distribution No.			1 / Distribution No.
Head Model #	Distribution No.	*"Order Frequency" (Days)*	Head Model #	Distribution No.	*"Order Frequency" (Days)*
IL30	2	0.5	PR20	1	1
DR30	1	1	PWR20	0.75	1.3
DL10	1	1	IR10	0.5	2
DL20	1	1	IR20	0.5	2
PWL10	0.5	2	IR30	0.5	2
PL10	0.5	2	IL10	0.25	4
DR20	0.5	2	IL20	0.25	4
DL30	0.5	2	DR10	0.25	4

WHAT IS THE NUMBER OF KANBAN NEEDED?

Having completed the Order Frequency Determination process, the Emca team is now ready to calculate the number of kanban for each golf club head. Before this calculation is made, a paradigm shift is required. Typically, inventory at Emca has been viewed in terms of quantities of parts. The team must now recognize that inventory takes on another form—time.

NOTE

The overriding principle of the lean organization is maximizing the effective use of time.

There are many equations available for Emca to use in determining the number of kanban needed in their kanban pull system. The one they selected is very flexible and appropriate for many environments. This kanban equation is expressed as follows:

$$\text{Number of Kanban} = \frac{\text{Average Daily Demand} \times (\text{Order Frequency} + \text{Lead Time} + \text{Safety Time})}{\text{Container Quantity}}$$

Let us define the various elements of the kanban equation.

Average Daily Demand: This is the current average quantity level of daily demand for a component. This is not a static number at Emca Golf, as seasonality exists in the demand for all of their products. Recalculating the number of kanban required is vital to the effectiveness of the kanban pull system as demand varies over time. As such, Emca Golf is going to document this recalculation procedure and include it as an element in their standard operating procedures.

Order Frequency: Order frequency is the frequency at which Emca's machining centers plan on setting up and running each golf club head. Order frequency is expressed in days, or a fraction thereof.

The order frequency is calculated by dividing 1 day by the number of set-ups per day allocated to the golf club head.

NOTE

Items with comparatively long order frequencies should be controlled by single use kanban cards. If these items were controlled by permanent cards, inventory of these items would be maintained in the supermarket and "sleep" for relatively long periods of time. In this case, the tool of production leveling should be explored to enable the consuming process to consume more frequently, but in smaller lot sizes.

Lead Time: Lead time is an estimate of how long the consuming process (assembly) will need to wait for an order of golf club heads once replenishment has been authorized. Otherwise stated, it is how much time elapses from the time a replenishment signal is sent to the supplying process until the consuming process is actually replenished. Factors that influence the duration of lead time include:

- The number of orders that arrived at the supplying process ahead of the one just sent.

- Machine cycle time.

- Quantity.

- Number and duration of set-ups.

- Replenishment signal method (manual or electronic).

- Product transit time.

NOTE

Order frequency should be at least equal to lead time in order to prevent replenishment signals getting crossed.

Lead time is an estimate because the variation in the factors we just reviewed can be significant from day to day and hour to hour. Lead time is expressed in days, or a fraction thereof.

Safety Time: Safety time is time (inventory) we allot to compensate for the impact of waste on the supplying process. This may take the form of machine downtime, response delay, absenteeism, scrap, shortages, and variation in demand-related issues. Safety time is expressed in days, or a fraction thereof.

Container Quantity: This is the number of units of each product that the team decided each container will hold. Emca Golf's "right-sized" containers' capacity for all golf club heads will be 10 units (as noted on the future state value stream map). "Less is frequently more" with container quantity because the material handler can easily transport smaller containers. The impact on the material handler of more frequent pick-ups and deliveries (the milk run path) when smaller containers are used must be considered as well.

Of all the elements of the kanban equation, container size is usually the one that the team will have the most freedom to change. This is beneficial because changing the container quantity is the most effective way to adjust the number of kanban, up or down, in the system without changing the level of inventory. In their initial calculation the team may finish with too many cards or too few. By changing the container quantity, they can get to a "workable" number of kanban.

When working through the kanban calculation process, the team must be certain they account for a waste factor only once. The impact of accounting for a factor more than once will be a greatly inflated level of inventory that could prove fatal to the kanban implementation process. A typical example of the opportunity to triple buffer is with demand. The team could include it as a safety stock factor, use the high end of the demand range, and consistently round up in the kanban calculation process.

The kanban equation described previously is embedded in the Kanban Calculation Worksheet that follows.

Emca Golf's kanban kaizen team will complete a Kanban Calculation Worksheet for each of the two machining resources. The team focused on Machine Center 1 first. This worksheet is where the team will tabulate and calculate the key numerical values associated with Emca Golf's kanban pull system.

Their first step is to enter the part number and average daily demand in the first two columns for the items that the team decided will be produced on Machine Center 1.

KANBAN CALCULATION WORKSHEET

RESOURCE: _____

Product	[(Average Daily Demand x (Order Frequency + Lead Time + Safety Time)] / Container Quantity = Number of Kanban

Note: Remember that order frequency should be at least equal to lead time in order to prevent signals from crossing.

KANBAN CALCULATION WORKSHEET							
RESOURCE:			Machine Center 1				
Product	[(Average Daily Demand	x (Order Frequency	+ Lead Time	+ Safety Time)]	/ Container Quantity	= Number of Kanban	
IL30	50						
DR30	40						
DL10	30						
DL20	30						
PWL10	29						
PL10	25						
DR20	20						
DL30	20						

The next step for the team is to enter the order frequency. In the case of the PWL10, every 2 days (see Order Frequency Determination Worksheet on page 36). For example, the order frequency for PWL10 is 1/0.5 set-ups/day = 2 days. Therefore, the plan is that every 2 days, the PWL10 head will be set up and manufactured on Machine Center 1.

KANBAN CALCULATION WORKSHEET							
RESOURCE:			Machine Center 1				
Product	[(Average Daily Demand	x (Order Frequency	+ Lead Time	+ Safety Time)]	/ Container Quantity	= Number of Kanban	
IL30	50	0.5					
DR30	40	1.0					
DL10	30	1.0					
DL20	30	1.0					
PWL10	29	2.0					
PL10	25	2.0					
DR20	20	2.0					
DL30	20	2.0					

The team's next steps are to determine the values for lead time, safety time, and container size. This information will be added to the Kanban Calculation Worksheet.

As we stated previously, lead time is an *estimate*. It is based on how long assembly will have to wait to be replenished. Remember, the elements of lead time include set-up time, processing time, signal delay, and transit time. And keep in mind that there may be other orders ahead of the signal just sent by assembly. The team estimated one day of lead time for all items to be produced on Machine Center 1.

KANBAN CALCULATION WORKSHEET						
RESOURCE:			Machine Center 1			
Product	[(Average Daily Demand	x (Order Frequency	+ Lead Time	+ Safety Time)]	/ Container Quantity	= Number of Kanban
IL30	50	1.0	1.0			
DR30	40	1.0	1.0			
DL10	30	1.0	1.0			
DL20	30	1.0	1.0			
PWL10	29	2.0	1.0			
PL10	25	2.0	1.0			
DR20	20	2.0	1.0			
DL30	20	2.0	1.0			

NOTE

As a general rule, order frequency should be at least equal to lead time in order to prevent replenishment signals from being crossed. In the case of the IL30 head, the team must manually adjust the "calculated" order frequency from 0.5 days to 1.0 days.

The team determined it needed 0.5 days of safety time to buffer demand spikes and other process-reliability problems.

As stated previously, the team decided its new standardized container quantity is going to be 10 units for all 16 club head part numbers.

Entering this data on the Kanban Calculation Worksheet for Machine Center 1, the team saw the following.

KANBAN CALCULATION WORKSHEET						
RESOURCE:			Machine Center 1			
Product	[(Average Daily Demand	x (Order Frequency	+ Lead Time	+ Safety Time)]	/ Container Quantity	= Number of Kanban
IL30	50	1.0	1.0	0.5	10	
DR30	40	1.0	1.0	0.5	10	
DL10	30	1.0	1.0	0.5	10	
DL20	30	1.0	1.0	0.5	10	
PWL10	29	2.0	1.0	0.5	10	
PL10	25	2.0	1.0	0.5	10	
DR20	20	2.0	1.0	0.5	10	
DL30	20	2.0	1.0	0.5	10	

The Emca team has quantified all five variables required by the kanban equation. Their next step is to work through the equation:

$$\text{Number of Kanban} = \frac{\text{Average Daily Demand} \times (\text{Order Frequency} + \text{Lead Time} + \text{Safety Time})}{\text{Container Quantity}}$$

The result of applying this equation to each item to be placed under kanban pull control at Machine Center 1 is as follows.

KANBAN CALCULATION WORKSHEET						
RESOURCE: Machine Center 1						
Product	[(Average Daily Demand	x (Order Frequency	+ Lead Time	+ Safety Time)]	/ Container Quantity	= Number of Kanban
IL30	50	1.0	1.0	0.5	10	13
DR30	40	1.0	1.0	0.5	10	10
DL10	30	1.0	1.0	0.5	10	8
DL20	30	1.0	1.0	0.5	10	8
PWL10	29	2.0	1.0	0.5	10	11
PL10	25	2.0	1.0	0.5	10	9
DR20	20	2.0	1.0	0.5	10	7
DL30	20	2.0	1.0	0.5	10	7

Note that we round up for the kanban calculation process because it is not physically possible to have a fraction of a kanban signal. Considering this is Emca's initial kanban implementation, it is also advisable to round up because the effect would be adding inventory to the value stream. Rounding down would remove inventory, thus making the value stream more sensitive.

EXERCISE

As a way to reinforce your understanding of the process just illustrated, refer to the CD file named "EMCA Calculation Worksheets-Exercise." Beginning with the worksheet tab labeled "Set-up Opportunities Calc.," enter the missing information into the highlighted cells. Then complete the worksheets labeled "Set-up Distribution Worksheet," "Order Frequency Worksheet," and "Kanban Calc. Worksheet MC2" (the separate kanban calculation worksheet for MC1 is already complete). Compare your solution with the file on the CD named "EMCA Calculation Worksheets-Solution."

The Kanban Calculation Worksheet below shows the results of the calculation process for Machine Center 2, based upon the Emca team's assumptions.

KANBAN CALCULATION WORKSHEET					
RESOURCE: Machine Center 2					
Product	[(Average Daily Demand x (Order Frequency	+ Lead Time	+ Safety Time)]	/ Container Quantity	= Number of Kanban
PR20	135	1.0	1.0	0.5	10	34
PWR20	111	1.3	1.0	0.5	10	32
IR10	100	2.0	1.0	0.5	10	35
IR20	100	2.0	1.0	0.5	10	35
IR30	100	2.0	1.0	0.5	10	35
IL10	75	4.0	1.0	0.5	10	42
IL20	75	4.0	1.0	0.5	10	42
DR10	60	4.0	1.0	0.5	10	33

NOTE

The kanban equation presented in this text possesses many interesting dynamics. For example, large container sizes drive down the number of kanban needed. Items that have long lead times require more kanban in the system, as do products from unreliable suppliers. As the number for kanban increases, so does the level of inventory. Inventory is the price paid for unresolved problems or waste.

The next calculation the team must perform determines the location of the order frequency cards on the kanban board. The location at which the team places these special cards establishes how many consumption signals (kanban cards) must accumulate to reach the authorization line on the kanban board. This calculation must be performed for all club heads controlled via kanban.

The order frequency card location is calculated as follows:

$$\text{Order Frequency Card Location} = \left[\frac{\text{Average Daily Demand} \times \text{Order Frequency}}{\text{Container Size}} \right] + 1$$

Let's look at the IL30 for an example.

Product	[(Average Daily Demand x (Order Frequency	+ Lead Time	+ Safety Time)]	/ Container Quantity	= Number of Kanban
IL30	50	1.0	1.0	0.5	10	13

The order frequency for the IL30 head is one day. The average daily demand for the IL30 is 50 heads. One kanban card will be attached to each container, and each container holds 10 heads.

$$\text{Order Frequency Card Location} = \left\lceil \frac{\text{Average Daily Demand} \times \text{Order Frequency}}{\text{Container Size}} \right\rceil + 1$$

Order Frequency Card Location for IL30 = $\lceil (50 \times 1) / 10 \rceil + 1$

Order Frequency Card Location for IL30 = 6

This means that the order frequency card location, indicated by a black checkered card, is the sixth card position under the authorization line on the kanban board. Five kanban cards must accumulate on the kanban board in order for the authorization line to be reached for the IL30 head.

ORDER FREQUENCY CARD LOCATION

	MACHINE CENTER 1		MACHINE CENTER 2
Head	# of Positions Under Authorization Line	Head	# of Positions Under Authorization Line
IL30	6	PR20	15
DR30	5	PWR20	16
DL10	4	IR10	21
DL20	4	IR20	21
PWL10	7	IR30	21
PL10	6	IL10	31
DR20	5	IL20	31
DL30	5	DR10	25

NOTE

Be aware of the effects of rounding during the order frequency card location calculation process. Rounding up means that you will be setting up an item less frequently, increasing capacity—as opposed to rounding down—setting up more frequently, thus reducing capacity. The impact of deciding to round down can significantly reduce capacity when applied to several products.

After the team completes the calculation process for Machine Centers 1 and 2, they check to see if the time value of the inventory seems to be correct. A good approach to testing the robustness of the team's solution is to simulate it in a training room. Here, the team can see, in advance, how "fault-tolerant" the solution is under circumstances such as extremes in demand or machine downtime.

WHAT IS A KANBAN FLOW DIAGRAM?

The purpose of developing a kanban flow diagram is to help the kanban kaizen team better visualize what the new process will actually look like and where, in the factory work area, key attributes of the kanban pull system will be physically placed. As such, the diagram should be drawn approximately to scale. It will also help in the process of generating the action items list necessary for the team members to "Go-Do." One technique in creating the kanban flow diagram is to create it on flip chart size paper using Post-its® to represent the key attributes.

The process for developing a kanban flow diagram is as follows:

1. Illustrate the supplying and consuming process.
2. Illustrate the kanban board position in the workplace.
3. Illustrate the supermarket position in the workplace.
4. Illustrate the milk run path and frequency.

WHAT'S NEXT AT EMCA?

Emca's activities have not gone unnoticed at Emca corporate. Today, Emca's other divisions are sending "scouts" to Emca Golf in order to benchmark them and learn their best practices. But Emca Golf still has a great deal to do to realize their "1st generation future state."

Near-Term Focus

Emca Golf's kanban team must now implement the kanban system they designed. This means they must:

- Make it real on the factory floor.

- Train those who will be using the system.

- Test the system.

- Adjust it as necessary.

- Properly maintain their new kanban pull system.

In doing so, they will further expand their knowledge of kanban pull systems and attain the many benefits they were seeking.

Mid-Term Focus

Emca Golf's mid-term focus will address the various kaizen burst opportunities remaining on their value stream map. Prioritizing which opportunities to attack first will require some speculation (also known as "what-iffing") as to the overall impact on the value stream.

Emca Golf needs to ask questions such as, Which improvement will have the greatest positive impact on the customer? Will it be if they:

- Conduct a machine reliability kaizen on Machine Center 1 and as a result, its uptime increases from 60% to 90%?

- Conduct a set-up reduction kaizen on Machine Center 2 that reduces the set-up time from 30 minutes to 8 minutes?

- Extend kanban to raw material suppliers?

- Extend flow to include packaging?

An important outcome of robust "what-iffing" with a value stream is gaining an understanding of the synergistic effect the next improvement has upon improvements already made. For example, if Emca decides that its next improvement focus would be to reduce the set-up time on Machine Center 2 and achieve a new changeover time of 8 minutes, the impact on the kanban pull system could be dramatic. Depending on how Emca decides to "spend" the new capacity they create, a complete kanban recalculation would be required, and the result would be an even greater reduction of replenishment lead times, and even lower lot sizes. By performing this "what-iffing" process, the value stream map truly guides the improvement process and ensures that improvements are "connected" to each other. This is how organizations avoid the "pocket of excellence" syndrome.

Long-Term Focus

Some surprises will occur at Emca Golf as they progress along their improvement journey. Their success with the initial lean implementation efforts will heighten an organization-wide sense of awareness about how much improvement potential really exists. They will be surprised to realize the magnitude of the improvement possible at Emca. They will understand that major improvement is possible on a scope broader than that illustrated on the future state value stream map for the golf club operation. This is typical of organizations poised to change their culture.

Emca will change their improvement perspective from "Lean Manufacturing" to "Lean Management," or from "doing lean" to "being lean."

Management at Emca Golf will truly "see the light." They will begin planning a lean function within Emca Golf that is staffed with full-time, certified lean managers.

They will create an annual improvement resource plan that offers the best mix of internal and external expertise. This will allow them to develop a broad in-house "improvement competency," one that is capable of effectively deploying a wide array of improvement tools and methods.

Emca Golf will link their improvement know-how and speed directly to the strategy of the organization. By doing, so they will leverage their new improvement competency as a competitive weapon in their markets.

Some of the advanced tools and skills they will learn and internalize include:

- How to apply lean to their product and process design functions.

- How to apply lean in non-manufacturing, administrative processes.

- How to integrate six sigma concepts with lean.

- How to improve the culture to the point where employees embrace continuous daily improvement as an aspect of their jobs. (For starters, every employee is expected to identify and rapidly implement 50 low-cost or no-cost improvements per year.)

- How to change management's role gradually to one of enabler and supporter of employee continuous improvement activity based on the scientific method.

A Bias for Action

A key improvement enabler for Emca is their ability to consistently demonstrate a bias for action. In the following chapter, we provide you with some tools to help you "Go-Do," so that you too can get to the action.

CHAPTER IV
Some Tools to
Help You "Go-Do"

❏ A suggested implementation strategy

❏ The kanban kaizen event

  – General and specific preparation check sheets

  – Event agenda

  – Event facilitation guidelines

  – Product data sheets

  – Suggested supplies checklist

  – Blank forms

A Suggested Implementation Strategy

Enterprisewide kanban pull system implementation first requires a well-thought-out strategy. The amount of time required to achieve full implementation of this strategy is a function of the scope of the specific value stream and the resources dedicated to the implementation process. The following are typical implementation strategy considerations and milestones.

Basic Precepts

- The implementation strategy is built upon previous improvements and is integrally linked to your future state vision.

- We encourage you to "start close to home." Develop an internal kanban pull systems implementation competency *before* taking the methodology to external suppliers.

- *If you need help—ask for it.*

Preparation Work

- Product family definition must be determined.
  - Many possible approach models for product family definition are available: by supplier, function, engineering content, processes, markets, etc.
- Value stream mapping (current state and future state) must be performed.

  — Collect key data points and verify where continuous flow is not possible. These areas are targeted for kanban pull system implementation.

- A requirements study must be performed.

- Evaluate the impact of the current state consuming process scheduling approach. Also, assess set-up times and process reliability impact.

Implementation Strategy Execution

- *Focus internally first.* Identify the internally produced products that should (and should not) be placed under kanban control. This determination will be the result of a balanced consideration of the demand volume, order frequency, and lead time of each product.

- Conduct kanban awareness training:
  - One-day key concepts workshops
  - Target audience
    - Client internal trainers
    - Process owners
    - IT support personnel
    - External suppliers

- Conduct kanban kaizen events (with your internal trainers):
  - Include in the training
    - Additional current state metrics
    - Requirements studies
    - Order frequency determination
    - The new pull system's design
    - New pull system prediction metrics
    - The new pull system's implementation
    - Test/fine tune the new pull system

- Integrate kanban into information system architecture.

- Perpetuate kanban kaizens with internal, certified training resources.

- Continually restate current and future state value stream maps as part of the continuous improvement journey.

- Deploy additional lean "tools" to further compress manufacturing lead time and reduce inventory.

External Focus

- Create a simple, but well documented, measurement system:
  - Establish a baseline of performance.
  - Share the measurements and the information on current performance with all suppliers.

- Conduct kanban pull system roll-out meeting(s) with suppliers:
  - Share your internal successes and misses with them.
  - Explain the resources available to help them.

- Identify the externally produced products that should (and should not) be placed under kanban control:

  – This will also establish the hierarchy of which suppliers you fold into the process first.

- Repeat the above-specified steps regarding kaizen, integration, perpetuation, and deployment.

A Few Kanban Kaizen Event Tips

- A qualified individual must facilitate the events.

- Define success up front.

- 50% improvement is good as a start.

- Focus on low-cost/no-cost solutions.

- The quality of the event planning determines the outcome.

- Do not try to establish too much change at one time — establish reasonable boundaries.

- Identify solid subteam leaders.

- Keep subteams staffed between five to seven people.

- Make sure each subteam has a member knowledgeable in the use of computer spreadsheet applications.

- Let event members know, in advance, that they are in for some hard work and long nights.

- Make sure that all the members work respectfully as a team without pulling rank.

- The more improvement ideas, the better for the company.

- Celebrate success with a vengeance!

- Pick a different area and repeat the process soon.

Pre-Event Preparation Checklist

- Is this event consistent with achieving the future state vision?

- Has a 5S level of 3 (demonstrated ability to sort, set in order, and shine)—minimum—been maintained in the processes/areas involved? This is important because if the areas involved don't have the discipline to maintain workplace organization, they will not have the discipline necessary to maintain a manual production control system.

- Is it clear how this event is linked to the future state value stream map? Does sufficient opportunity exist to improve the key metrics and transfer knowledge?

- Do any major issues exist that threaten a positive outcome (i.e., production or change resource shortages, conflicting activities, unwillingness to disrupt)?

- Have you publicized the event for awareness to the entire plant population?

- Have you drafted a preliminary agenda?

- Does a *positive spirit* exist in the attitude of supervisors and workers?

- Has management sponsored and/or is it willing to participate in the event?

- Has a discussion occurred with management regarding expectations and likely outcomes? Are you aligned?

- Are the area workers and supervisors clear on how their jobs and responsibilities may be impacted?

- Have you communicated a detailed agenda with clearly defined boundaries?

- Have you identified the team membership: multishifts, internal, or external suppliers/customers?

- Have you determined the kaizen team membership and limited total team size to 20?

- Have you anticipated subteams and identified subteam leaders?

- Have you conducted a team members' and supervisors' meeting, in which the details of the event were addressed?

- Have you arranged a meeting room for the entire kaizen team and an additional breakout room for subteams?

- Have food, beverages, and audio-visual equipment been supplied?

- Have you determined the supplying/consuming processes to be linked via kanban? Limit to 25 part numbers per subteam of five members. Consider:
  - Chronic shortages.
  - Chronic overages.
  - A-B-C-D analysis findings.
  - "Dead" inventory items.

- Have you mitigated management expectations that inventory levels will always decrease for all items placed under kanban control?

- Have you performed pre-event communications to all impacted?

- Have you completed *Product Data Sheets* for all items targeted for placement under kanban control?

- Have you ordered and verified that all the necessary materials and supplies are available (see the *Supplies Checksheet*)?

SUGGESTED KANBAN KAIZEN EVENT AGENDA

Day 1 A.M.:

- Management kick-off
- Kanban pull system training

Day 1 P.M.:

- Kanban pull system training (continued)
- Subteam formation (based on the number of focus areas)

Day 2 A.M.:

- Current state process definition
- Review and verify Product Data Sheets (pages 61-62)
- Complete Product Data Sheets as necessary (this should be minimal)

Day 2 P.M.:

- Perform a requirements study and order frequency determination
- Complete Kanban Calculation Worksheets
- "Verify" Kanban Calculation Worksheets
- Future state visioning
- Develop Kanban Pull System Flow Diagram
- Action item definition

Day 3 A.M. & P.M.:

- Action item execution (i.e., design and build boards, cards; build bin holding systems, etc.)
- Draft a new standard operating procedure (SOP)
- Test the new kanban pull system
- Refine the system as necessary
- Update SOP as necessary

Day 4 A.M.:

- Train all necessary in SOP

- ID open action items and completion dates

- Develop team report-out presentations

Day 4 P.M.:

- Final team report outs, and celebration

KANBAN KAIZEN EVENT SUB-TEAM LEADER CHECKSHEET:

- Review and verify that the product data sheets for the products to be placed under kanban control are complete and correct.

- Lead the data collection process for any missing information your team needs.

- Lead the calculation process with your team and resolve any issues that are subjective in nature.

- Lead the new process visioning/brainstorming discussions.

- Design on paper the Kanban Flow Diagram. Be sure to indicate:
  - Board positioning
  - Bin and line stock positioning
  - Card and product paths

- List the action items necessary to get the paper design to become reality.

- Lead the interim management report out.

- Lead the execution of the identified action items.

- Lead the creation of the new standard operating procedure.

- Test the new process and make any adjustments as necessary.

- Train all necessary in the new SOP.

- Identify open action items and assign people responsible for closure.

- Lead the development of the final report out.

- Lead the follow-up process regarding open action items.

BLANK FORMS

Kanban Pull System Product Data Sheet

Date: _____   Sheet : _____ of _____

Product Name: _____   Storage Location(s): _____

Product ID Number: _____   Current Replenishment Method: _____

Family/Comm. Group: _____   A-B-C-D Designation: _____

Number of Customers: _____

Product Inventory Data:

Storage Qty.: _____   Current Inventory:

Qty.: _____

Container Qty.: _____   DOS: _____

Storage Method(s): _____   12-Month Inventory Profile: _____
(Provide a graphical distribution)

Planned Safety Stock: Current Min./Max. Controls. (as applicable)

Qty.: _____   Min. _____

DOS: _____   Max. _____

Consumption Data:

12 Month Historical

Monthly Consumption distribution: (Provide this distribution) _____

Is consumption cyclical?   ☐ Y   ☐ N

If yes, describe nature: _____

Order Qty. Distribution: (Provide this distribution) _____

Average: _____

Range: _____

Order Frequency: _____

Average: _____

"Drop-In" Sensitive?   ☐ Y   ☐ N

Frequency: _____

Qty. per Occurrence: _____

Future

Existing Firm Order Data: (provide this information) _____

Existing Forecast Order Data: (provide this information) _____

Factors Likely to Significantly Alter Next 12 Mo. Demand: _____

Obsolescence?   ☐ Y   ☐ N

Other: (list) _____

Current Daily Demand Determination: _____

(Continued on next page)

61

Kanban Pull System
Product Data Sheet, *continued*

Date: _____   Sheet : _____ of _____

Supplier Data:

Supplier Name: (internal or external)   _____

Total # of P/N's Produced By this Supplier: _____

    List along w/A-B-C-D Designator: (provide this data)   _____

Current Process Cycle-time: _____

Typical Lot Size(s) and Range: _____

Changeover Time

    Average (typical):   _____

    Range: _____

Current Run Frequency

    Average (typical):   _____

    Range: _____

Shortage History (12-Mo.)

    # of Occurrences: _____

    Qty. per Occurrence: _____

    Date of Each Occurrence: _____

    Reason for Each Occurrence: _____

Quality Defect Rate: _____

Current Capacity: _____

    # of shifts/hrs. available: _____

    # of shifts/hrs. operational: _____

Key Equipment Data

    Current Reliability

        MTBF   _____

        MTTR   _____

        OEE   _____

        Constraint: _____

Kanban System Kaizen Event
Suggested Supplies Checksheet

QUANTITY	ITEM
1	Training room with (9) 3′ × 6′ tables, chairs, projector: O/H & LCD, screen, VCR
1 per team	Flip charts & stands
1 per team	Team break out areas
1 per team	Laminating machine per team
1 set	Laminating supplies (enough for 300, 8.5″ x 11″ sheets)
3 per team	White board, Hunt® display board #951450 or equivalent (40″ x 60″)
4 each	20′ roll of plastic coated wire
10 rolls	Plastic tape: black, 1/4″ wide
500 each	Plastic sleeves to hold kanban cards (packing slip holders work nicely)
3 each	Single-hole punch
1 each	Paper: ream, white 8.5″ x 11″
100 each	Card stock: white, red, black, 8.5″ x 11″
1 per team	Screwdrivers and Phillip's head screws
1 per team	Hand pliers
1 per team	Hand wire cutter
1 per team	Paper shear
1 each	Velcro®: roll, self-adhesive, 1″ wide
30 pkg.	Adhesive 1-1/4″ adhesive clips
5 pkg. each	Stick-on letters and numbers, 2″: Black, red, yellow, green
1 unit	Label maker: Handy-mark by Brady® (or equivalent)
1 set	Label maker supply kit: mix of colors and widths
1 per team	Packaging tape: clear
1 per team	Yardsticks
1 per team	Scissors
Teams will need regular access to:	
1 per team	Color printers
1 per team	PC's with spreadsheet software installed with autosave add-in functional
As checked off, include:	
	Materials to build parts presentation racks (tubing or wood)
	Right-sized containers
	Timecard rack
	Digital camera

REQUIREMENTS STUDY FORMS

RESOURCE CYCLE TIME ANALYSIS WORKSHEET		
Component Part Number	Resource #1 Cycle Time	Resource #2 Cycle Time

MACHINE ALLOCATION TABLE			
Component Part Number	A-B-C-D	Resource #1	Resource #2

REQUIREMENTS STUDY WORKSHEET

COMPONENT: _____

Component Part Number	Average Daily Demand	A-B-C-D Designation	Resource #1 Cycle Time	Resource #2 Cycle Time	Selected Resource ID	Resource #1 Loading	Resource #2 Loading
					TOTALS:		

ORDER FREQUENCY DETERMINATION FORMS

EQUIPMENT ANALYSIS WORKSHEET							
Resource ID	Capacity per Shift (sec.)	Number of Shifts	Gross Machine Capacity (sec.)	Uptime %	Net Machining Capacity (sec.)	Set-Up Time (minutes)	Set-Up Time (sec.)
Resource # 1							
Resource # 2							
				TOTALS:			

SET-UP OPPORTUNITIES CALCULATIONS			
Resource ID	Net Capacity —	Resource Loading Time =	Time Available to Set up
Resource # 1			
Resource # 2			

Resource ID	Time Available for Set Up /	Set up Time =	Number of Set-up Opportunities/Day
Resource # 1			
Resource # 2			

SET-UP DISTRIBUTION WORKSHEET					
MACHINE CENTER 1			MACHINE CENTER 2		
Component Part Number	Average Daily Demand	Distribution	Component Part Number	Average Daily Demand	Distribution
Total Allocations			Total Allocations		
Total Opportunities			Total Opportunities		
Discretionary			Discretionary		

ORDER FREQUENCY DETERMINATION WORKSHEET

Resource # 1

Component Part Number	Distribution No.	1 / Distribution No. = "Order Frequency" (days)
Total No. Distributed		
Flex Capacity Time (sec.)		

Resource # 2

Component Part Number	Distribution No.	1 / Distribution No. = "Order Frequency" (days)
Total No. Distributed		
Flex Capacity Time (sec.)		

KANBAN CALCULATION FORMS

KANBAN CALCULATION WORKSHEET					
RESOURCE: _____					
Component Part Number	[(Average Daily Demand x (Order Frequency + Lead Time + Safety Time)]	/ Container Quantity	= Number of Kanban		

Note: Remember that order frequency should be at least equal to lead time in order to prevent signals from crossing.

ORDER FREQUENCY CARD LOCATION WORKSHEET				
RESOURCE # 1			RESOURCE # 2	
Component Part Number	# of Positions Under Authorization Line		Component Part Number	# of Positions Under Authorization Line

CHAPTER V
Managing the System

MANAGING THE MIXED MODEL KANBAN SYSTEM

Now that the kanban system is physically designed and in place, Emca must develop and communicate a plan to manage the system on a daily basis. The plan must cover who is responsible for the various aspects of managing the system, and it must define normal and abnormal conditions so that the system users can respond quickly and appropriately if abnormal conditions occur.

Because kanban is a production control function engineered into the value stream, the value stream manager is ultimately responsible for the health, well-being and continual improvement of the mixed model kanban system. It is the value stream manager that must make sure everyone adheres to the documented procedures for operating the system. The value stream manager will require support from the various kanban system users to help him or her be successful. First and foremost, it is through the proper and diligent use of the kanban system that production control, material handlers, and supplying and consuming process owners provide the best support. System users can provide an additional level of support by giving the value stream manager real-time kanban system performance feedback, especially regarding the abnormal conditions discussed in this section, and by sharing additional improvement ideas.

Under normal conditions, the production control manager, with frequent involvement of the value stream manager, will need to manage the day-to-day operational situations that commonly arise. Such situations include:

- Introducing single use kanban cards
- Tracking and adjusting kanban equation variables
- Introducing new products into the kanban system
- Removing obsolete products from the kanban system
- Transitioning from permanent to single use kanban cards
- Transitioning from single use to permanent kanban cards

Every once in a while, an abnormal condition will present itself. When this happens, it is critical that the kanban system users identify the condition as soon as it occurs and immediately bring it to the attention of either the value stream manager or the production control supervisor. Abnormal conditions include but are not limited to:

- All items achieve the authorization line at the same time

- No items achieve the authorization line for an extended period of time

- Kanban cards accumulate above the authorization line

- Lost kanban cards

- Excessive delays in material handling (container and kanban card movement)

What follows are some methods on how to respond properly to these "normal" and "abnormal" conditions.

MANAGING NORMAL CONDITIONS

Introducing Single Use Kanban Cards

Although it was not the case with Emca Golf, it is common that most of the part numbers produced at a supplying resource and controlled using a mixed model kanban system are actually controlled through single use kanban cards. The 80/20 rule supports this: 80% of total unit demand is typically derived from 20% of the products manufactured. The fact that most products can be controlled by single use kanban cards is good news because it makes controlling and managing the mixed model kanban system much easier. As discussed earlier, single use kanban cards are used to control the "D" or special order items. "D" items are consumed so infrequently that it does not make sense to always maintain inventory for them in the supermarket (as would be the case if we controlled them using permanent cards). We know the capacity exists at the supplying process to produce these "D" items, because an element of performing the order frequency determination process was to estimate how much flexible capacity should be set aside to set up and produce them. When the consuming process signals that one of these special order items is needed, it is typically the responsibility of production control to place the single use kanban card on the kanban board in the column labeled "single use."

Machine Center 1 Kanban Board

Single Use	Past Due	IL30	DR30		DL20	PWL 10	PL10	DR20	DL30

Emca production control needs to think carefully about the timing of placing the single use kanban card on the kanban board. Factors to consider include:

1. The lead-time impact to the other consuming processes of permanent card products produced by the supplying resource, and

2. The required delivery date/time of the single use product to the consuming process.

Tracking and adjusting kanban equation variables

Emca Golf production control is primarily responsible for maintaining the integrity of the kanban calculation process.

Number of Kanban = Average Daily Demand x (Order Frequency + Lead Time + Safety Time) / Container Quantity

Demand: Keeping a careful eye on the current demand for each item and its demand pattern is a vital task that Emca production control must perform in properly managing their mixed model kanban system. Factors that influence the nature of the demand pattern include seasonality and where the product is in the lifecycle. These factors need to be well understood. As significant shifts in demand occur, the order frequency and kanban calculation processes will need to be updated by Emca production control.

Order Frequency: Adjusting order frequencies is a relatively common activity. Very few elements of manufacturing are static, especially in an organization like Emca Golf, which is aggressively pursuing continuous improvement. The most common drivers of the need to adjust order frequency include changing the number of items controlled by kanban, factory capacity changes such as changes in shift structure or the number of supplying process resources/machines, and lean improvements such as improving set-up times and improving machine reliability. Staying on top of these changes is vital. They require Emca to update the order frequency determination, which in turn requires them to recalculate the number of kanban required. If Emca were to fail to do so, they would be placing the value stream in jeopardy.

Lead time: The most common reasons requiring modification of the lead time element of the kanban equation are changes occurring in the lot sizes of the products produced at the supplying resource, cycle time improvements at the supplying resource, and changes in the number of products the supplying resource is responsible for producing. As any of these changes occur at Emca Golf head fabrication, production control must recalculate the number of kanban required in the value stream and either add or remove kanban as appropriate.

Safety time: Common reasons to adjust the safety time element of the equation relate to changes in the reliability of the supplying process controlled with kanban and changes in the stability of average daily demand. For example, as reliability improvements occur at the supplier of the golf club head raw material, less safety time inventory is necessary. The reduction in safety time inventory necessitates a recalculation of the number of kanban required.

Container Quantity: Altering container quantity is a common occurrence. Most frequently the container quantities become smaller as organizations reduce inventory levels and progress toward the ideal value stream state of continuous one-piece flow. Altering container quantities is also the most convenient way of keeping the number of kanban in the value stream at a manageable level.

> **NOTE**
>
> *It is important that Emca keep an accurate history of all changes made to the kanban system so as to facilitate future problem solving and improve organizational understanding of mixed model kanban system functionality.*

Introducing New Products Into the Kanban System

As new products are introduced in the marketplace, additional production requirements will be placed on the supplying process. Assuming that the new products are designated "A" or "B" items, Emca Golf production control must completely redo the requirements study, the order frequency determination, and the kanban calculation process. Kanban cards must be created, the kanban board must be modified, space in the supermarket must be created, and inventory for the new items must be positioned in the supermarket.

Removing Obsolete Products From the Kanban System

If Emca Golf were to declare products obsolete (assuming no "spares" requirements exist), the supplying process would obviously no longer have to produce them. If these obsolete products were controlled with single use kanban cards, then Emca production control would simply not have to worry about issuing any more of them to the supplying process. If the obsolete products were controlled by permanent cards, then the following activities must occur.

1. The kanban board and supermarket must be modified to exclude references to the obsolete items.

2. A new order frequency determination must be performed on the balance of the parts controlled by kanban at the supplying resource, because the capacity of the supplying resource is affected by the obsolescence.

3. The number of kanban must be recalculated along with the order frequency card location.

Transitioning From Permanent to Single Use Kanban Cards

If the demand level of a product produced at Emca Golf was to decrease to the point where it no longer made sense to control the product with permanent cards, then production control would have to completely redo the demand requirements study, the order frequency determination, and the kanban calculation process. Additionally, the inventory in the supermarket that traditionally exhibited permanent cards must now exhibit single use cards. The inventory in the supermarket must be depleted, which will free space in the supermarket for new products, and the kanban board must be modified to exclude reference to the item.

Transitioning From Single Use to Permanent Kanban Cards

If, on the other hand, the demand level of a product produced at Emca Golf was to increase to the point where it now made sense to control the product with permanent cards instead of a single use card, then production control would have to completely redo the requirements study, the order frequency determination, and the kanban calculation process. Additionally, inventory for the item needs to be created and placed in the supermarket exhibiting a permanent card. Space would need to be created in the supermarket for the new permanent product, and the kanban board would need to be modified to provide a designated column for the item.

Managing Abnormal Conditions

All Items Achieve the Authorization Line at the Same Time

When this condition occurs, the mixed model kanban system is telling the users it needs more capacity. The quickest and simplest way to create this capacity is to lower the position of the order frequency cards on the kanban board. The capacity you gain will come from setting up products less frequently than before. The penalty to be paid for gaining this capacity is longer lead times because the lot sizes will increase.

No Items Achieve the Authorization Line for an Extended Period of Time

When the system users observe that the authorization line is not being achieved for any products for an extended period of time, issues related to demand exist. The situation may be caused by disruptions in consumption by the consuming process. Possibly, their line is down. Another possible cause is that the average daily demand numbers may be significantly overstated. In the case of the latter, production control must completely redo the requirements study, the order frequency determination, and the kanban calculation process. Changes in the position of the order frequency card may also be necessary.

Kanban Cards Accumulate Above the Authorization Line

When kanban cards accumulate above the authorization line, either the supplying process is not producing and delivering within the estimated lead time, or the consuming process is consuming at a rate significantly higher than originally planned. In the case of the former, the root cause typically turns out to be an inaccurate estimation of the average number of orders in the FIFO queue at the supplying process. If the root cause turns out to be the latter then production control must redo the requirements study, the order frequency determination, and the kanban calculation process. Changes in the position of the order frequency card may also be necessary.

Lost Kanban Cards

Production control must perform a root cause analysis as to why kanban cards are being lost. In most cases, it turns out to be either a discipline or training issue. In both instances the opportunity exists to mistake-proof the methods of kanban card attachment. Additionally, changes to the frequency of auditing cards may be necessary in the short term.

Excessive Delays in Material Handling (Container and Kanban Card Movement)

Production control must perform a root cause analysis as to why material handling delays are excessive. In most cases, it turns out to be either a discipline or training issue. In some cases the need to develop precise material handling schedules with assigned responsibility may be necessary to ensure timely kanban card and container movement.

CHAPTER VI

Frequently Asked Questions

WHAT HAPPENED TO ALL THE INVENTORY?

All shopfloor operators will need to be trained so they do not panic when they no longer see the oceans of inventory in the workplace. The operators on the implementation team are the means of communication to the workforce on this issue.

WHAT ABOUT THE PIECEWORK INCENTIVES CURRENTLY IN PLACE?

If the violation of kanban pull rules and overproduction are to be prevented, piecework mentality and systems must be abolished. Meeting takt time and the level of multiskill capabilities of your employees are now your new priorities.

HOW WILL THE SYSTEM BE MANAGED?

Your new kanban-based production control system must be documented. You must, at a minimum, clearly explain:

1. How the entire system is designed to work.

2. Who is responsible for each task within the system.

3. Who is responsible for managing the system.

4. Who will audit the system, and how frequently that audit will occur.

5. How frequently the new kanban pull system will be re-evaluated for necessary adjustments.

6. The standards to be met in the system (i.e. card design, kanban board design, container color coding).

7. Appropriate graphical flow diagrams.

8. An explanation of how the system's users will be trained in their role to use the system.

WHAT ABOUT "ALLOCATION"?

Allocation is an issue that often presents itself early in kanban pull system implementation. Allocation is a process in which material is electronically marked as sold in the computer system, but is not physically removed and shipped. What happens to the kanban card? Our advice is to NOT hold the cards "hostage" by delaying their placement on the kanban board. When the product finally ships, the supplying process will experience huge demand spikes. If you are certain that that the material will actually ship, the material should be removed from its rack and placed in a waiting-to-ship area. The kanban cards should be placed on the kanban board. A conscious effort should be made to compress the allocation time frame and reduce the manufacturing lead time for the product, thus possibly negating the need to allocate altogether. Efforts to further level the customer demand should also be attempted.

WHAT ROLE DOES MRP PLAY?

A long-range demand, materials, and capacity planning capability is still needed and MRP can assist with this. MRP can no longer play a role in the daily execution of production control for items placed under kanban control. That is the job of the kanban pull system.

WHAT WOULD HAPPEN IF WE TRIED TO RUN THE KANBAN PULL SYSTEM WITHOUT "PRIMING" THE SYSTEM FIRST?

Do not try to run the system without priming it first or try to run the system from an unsteady state. This is one of the most frequently made mistakes when implementing kanban. At best, it will lead to difficulty in managing the kanban pull system, and at worst it could lead to the failure of the system. An unprimed kanban pull system will not work properly because while you are building inventory for the supermarket, you may not be responding to the immediate needs of the consuming process. Take the time to build and properly position the needed inventory. Do not jeopardize the kanban pull system implementation process by creating "artificial" shortages at the consuming process.

How often should you resize a kanban pull system?

The system must be resized as changes in consumption and supply variables occur. Also, in the early stages of a product being placed under kanban control, we encourage you to verify the kanban sizing more frequently. Additionally, feedback from the users of the kanban pull system may indicate kanban resizing is necessary. Neglecting to resize the system is a frequent cause of failure.

What happens if I lose kanban cards?

The system is at risk of failing. By losing kanban cards, we can lose visibility of two very important things. First, if the card was on a container when it was lost, we no longer know if the material the card was attached to is authorized. Secondly, if the kanban card was lost after the consuming process emptied the container, the supplying process will be unaware that consumption occurred. Therefore, replenishment will be delayed because the authorization point will be reached that much later. This increases the risk of a shortage occurring at the consuming process.

Why not just forget the kanban cards and board altogether and just use painted squares on the floor, or just empty containers?

By using the "kanban square on the floor" technique, authorization-to-produce priority will be difficult to maintain in a high-mix environment. It also assumes the consuming and supplying processes are in the line of sight of one another.

How will I know if I have lost kanban cards?

By periodically auditing (counting) the number of cards in the system, you will discover if you have lost any cards.

DOES A MANUAL KANBAN PULL SYSTEM DISCONNECT INVENTORY DATA FROM MY IT SYSTEMS?

Yes. The inventory data is kept with the inventory via the kanban card.

WHAT ABOUT "ELECTRONIC KANBAN"?

We generally do not recommend using electronic kanban, especially at the early stages of implementation. We say this for several reasons. First, most of the "solutions" we have come across are nothing but a repackaging of "old school" reorder-point logic. As of the writing of this workbook, we have heard rumors of some software vendors attempting to "get it right," but they are still working on it. Secondly, without the user understanding the logic of kanban pull (by being involved with implementing it manually first), you would just be throwing another piece of software at them. Also, all electronic kanban pull systems are dependent on accurate data input and timely computer transactions. This can be problematic, as data integrity haunts most companies. In addition, unless you are in the software development business, computer transactions are non-value-added by definition.

WHO IS RESPONSIBLE FOR MOVING THE EMPTY AND FULL CONTAINERS AND CARDS, THE MILK RUN PATH?

We strongly suggest that the material handler or water beetle perform this so your value adders can focus on adding value. These people perform a vital role in the kanban pull process. In many companies, their roles and images change as a result.

WHAT IS DIFFERENT WHEN IT COMES TIME TO EXTEND KANBAN PULL TO EXTERNAL SUPPLIERS?

Not that much will really change. The signaling method may have to change (the use of web cams may help with this). All of the variables in the kanban equation still apply, but they need to be considered in a different context. Your supplier will need to tell you what their order frequency is. No, you cannot determine it for them! The lead time is likely to be longer, and safety time issues may be different.

Several other factors are typically re-evaluated while extending kanban externally. They include supply base consolidation initiatives, improving the sharing of accurate, long-term forecast data, and a better understanding of the suppliers' shipping logistics. Of course, it helps to go into it with a positive relationship based on mutual benefit and trust, rather than the "rammer-jammer" mentality to working with your supply base.

HOW DO I MANAGE THE SITUATION WHERE THE SUPPLYING RESOURCE I PLACE UNDER KANBAN PULL CONTROL MUST ALSO PRODUCE PRODUCTS FOR OTHER CONSUMING PROCESSES THAT ARE NOT IN MY AREA OF KANBAN IMPLEMENTATION RIGHT NOW?

In this instance, at the order frequency determination stage, you must set aside a certain amount of capacity at the supplying resource for the production of these articles.

WHAT ABOUT THE "GOLDEN ROPE"?

The term "Golden Rope" was coined at Toyota. After a kanban kaizen event, the situation may arise where the existing inventory level in the supermarket may be very large compared to the inventory authorized by the kanban pull system. This condition may prevent the permanent cards from circulating, thus preventing the system from being tested as designed. The Golden Rope concept directs assigning single use kanban card(s) to the excess inventory. Then the inventory should be removed from the supermarket and placed in a separate area behind a "golden rope." The inventory behind the golden rope may be gradually consumed after proving the system works with permanent kanban cards as designed.

CONCLUSION

As is the case with most worthwhile endeavors, implementing kanban pull will not be easy. However, understanding the principles of kanban, having a reliable implementation method to follow, and anticipating the challenges ahead makes the improvement effort much easier. We refer to this implementation method as "reliable" because when you follow the methodology, it works. Conversely, when you do not follow the methodology or skip steps, it does not work. So, do not skip any steps.

The task of trying to run and manage parallel production control systems—the old system and the new kanban pull system—will wear you down over time. After you have conducted your initial pilot implementation, we challenge you to aggressively implement kanban pull wherever it is appropriate in your value stream. Do not underestimate the magnitude of the task. If you need help, do not wait long to ask for it. Good luck!

GLOSSARY

Batch and queue: refers to the usual movement of part lots in mass-production practices. Typically, large lots of a part are made and sent as a batch to wait in queue for the next operation in the production process. *Contrast with one-piece flow.*

Cell: a logical, efficient, and usually physically self-contained arrangement of machinery, tooling, and personnel to complete a production sequence. The cell enables one-piece flow and multiprocess handling.

Cellular manufacturing: manufacturing by the use of cells. See *Cell.*

Champion: an individual, from any level of the organization, who has the authority and responsibility to inform, support, and direct a team effort to implement and integrate a new tool, method, technique or technology, etc. The champion is a first-line resource for all the participants and, in some cases, has the authority to allocate the organization's resources during the life of the project. Also called lean champion or project champion.

Changeover: altering a process to accommodate a different product model.

Changeover time: the time between the last good piece off one production run and the first good piece off the next run, producing at the target volume.

CNC machine: See *Computer numerical control machine.*

Computer numerical control (CNC) machine: a versatile and sophisticated machine used in manufacturing for its complex motion-control capabilities, which offer improved automation, consistent and accurate workpieces, and flexibility.

Continuous flow: see *One-piece flow.*

Cycle time: specifically, the time that elapses from the beginning of one operation or one part of a process until its completion. Operator cycle time is the total time for an operator to complete one cycle of an operation, including walking, loading, unloading, inspecting, etc. Machine cycle time is the time between when the "on button" is pressed until the machine returns to its original position after completing the operation.

Defect: nonconformance in a product or part, or departure of quality from the intended effect. In mistake-proofing terminology, a defect is not the same as an error. A defect is the result of an error.

Discrete products: products typically made by the assembly of a component parts.

Downtime: manufacturing time that is not useable because of equipment problems, lack of materials, lack of necessary information, or operator unavailability.

Five S (5S): an improvement process, originally summarized by five Japanese words beginning with S, to create a workplace that will meet the criteria of visual control and lean production. *Seiri* (sort) means to separate needed tools, parts, and instructions from the unneeded and to remove the latter. *Seiton* (set in order) means to neatly arrange and identify parts and tools for ease of use. *Seiso* (shine) means to clean and inspect. *Seiketsu* (standardize) means to require as the norm that everyone sort, set in order, and shine at frequent (daily) intervals to keep the workplace in perfect condition, and also to make use of visual control systems. *Shitsuke* (sustain) means to maintain the five S gains by training and encouraging workers to form the habit of always following the first four S's. Also called workplace organization and standardization and referred to as the five pillars of the visual workplace. (Safety concerns are sometimes added to the process and referred to as the sixth S.)

Flexibility: the ability to rapidly respond to unforeseen circumstances.

Flow: the progressive achievement of tasks as a product proceeds along the value stream, including design to launch, order to delivery, and raw materials into the hands of the customer without stoppages, scrap, or backflows. Flow can apply to the movement of information as well as material.

Just-in-time: the first of the two major pillars of the Toyota Production System (the second being autonomation), just-in-time is a system for producing and delivering the right items to the right place at the right time in the right amounts, eliminating buffer inventories. This technique approaches just-in-time when upstream activities occur minutes or seconds before downstream activities, so that one-piece flow is possible. The key elements of just-in-time are flow, pull, standard work (with standard work-in-process inventories), and takt time.

Kaizen: composed of the Japanese *kai*, meaning "to take apart," and *zen*, meaning "to make good." Kaizen is the gradual, incremental, and continual "improvement" of activities so as to create more value and less non-value-adding waste. Its success depends on the total commitment of the work force to increasing efficiency and reducing costs.

Kaizen event: a planned and structured event that enables a group of associates to improve some aspect of their business.

Kanban: meaning "signboard" or "signal" in Japanese. In the context of production, it refers to visual production control signals.

Kanban system: in the context of production, it refers to a visual production control system that signals the need to replenish the supplying processes.

Lead time: the total amount of time required to get an order to the customer.

Lot: a volume of product that has been produced as a batch.

Manufacturing lead time: the amount of manufacturing time taken from the issuance of raw material through the production process to the completion of the saleable product.

Material requirements planning (MRP): a computerized system used to determine the quantity and timing of the supply of materials used in a production operation. MRP systems include a master production schedule, a bill of materials specifying each item needed, and information about current inventories from which to schedule the production and delivery of needed items.

Material handler: a person on the production floor who paces the entire value stream to ensure that integrity is maintained by the timely transporting of material, containers, and kanban signals.

Mean time between failures (MTBF): a rating that indicates the average ability of an item or system to perform a required function, under stated conditions, without failure, for a stated period of time. It is determined by dividing the time frame being analyzed by the number of breakdowns. It is a reliability rating.

Mean time to repair (MTTR): a rating that indicates the average time (rapidity and ease) in which maintenance operations can be performed to either prevent malfunctions or to correct them if they occur. It is determined by dividing the total downtime for repairs by the number of repair incidents. It is a maintainability rating.

Milk run: the routing of a supply or delivery vehicle to make multiple pick-ups or drop-offs at different locations. The route of a material handler within a factory is called a milk run.

Mixed-model: a value stream that is designed to accommodate multiple product models.

MTBF: See *Mean time between failures.*

MTTR: See *Mean time to repair.*

One-piece flow: the manufacturing process in which product flows without waiting through all necessary operations, one piece at a time, without back-flows or excess inventory. This is also called single-piece flow. *Contrast with batch and queue.*

Operation: an activity or activities performed on a product by a single machine or person. *Contrast with process.*

Order frequency: the frequency at which the consuming process will place orders to the supplying process for the production of a component or product.

Order frequency determination: a method of establishing the net capability of a resource to produce. Also quantifies the time available at each machine center for set-up.

Paradigm: a closely held perception of reality, frequently unquestioned and difficult to change, that conditions all our thinking about and our understanding of the world or some aspect of experience.

Pilot: an experimental task or exercise to determine the viability of a concept.

Process: a sequence of operations (consisting of people, machines, materials, and methods) for the design, manufacture, and delivery of a product or service. *Contrast with operation.*

Pull: a system of production and delivery instructions in which replenishment does not occur absent a consumption signal by the downstream customer.

Push: conventional production, in which production schedules are pushed along based on sales projections and availability of materials. It leads production employees to make as much product as they can as fast as they can, even if the next process is not ready to use the materials, which causes large work-in-process inventories. *Contrast with pull.*

Queue: an accumulation of inventory authorized by a push signal.

Requirements study: the quantification and study of the amount of resource consumption necessary to satisfy the needs of a consuming process.

Safety time: the time (inventory) allotted to compensate for the impact of waste on the supplying process.

Set-up time: the time between the last good piece off one production run and the first good piece off the next run, producing at the target volume.

Signboard: the English translation of the Japanese word "kanban." See *Kanban*.

Supermarket: a storage location for inventory authorized by a kanban pull system.

System: a set or an arrangement of things so closely related or connected as to form a unit or organic whole. From system dynamics we learn that systems have typical "behavior" patterns and feedback loops that are both positive and negative. Typical patterns of complex systems include resistance, drift to low performance, general parameter insensitivity, heightened sensitivity to particular influence points, and conflict between long-term and short-term response. Systems also exhibit patterns of growth, decline, oscillation, equilibrium-seeking, and goal-seeking.

Takt time: the rate at which product must be turned out to satisfy market demand. It is determined by dividing the available production time by the customer demand.

Team: a group of people who rely on cooperation, trust, and communication to achieve a common set of objectives or targets. A cross-functional team is made up of people from different departments in an organization.

Toyota Production System (TPS): a manufacturing efficiency model built upon three key factors: reduced lot sizes to allow for production flexibility, the control of production parts so that parts are always available when and where they are needed, and the arrangement of production equipment in logical order of assembly.

TPS: See *Toyota Production System*.

Uptime percent: the percent time a resource is actually available for production. The formula to determine uptime is: (net available resource time ÷ gross available resource time) × 100.

Value stream: all the activities (both value-added and non-value-added) required within one company to design and provide a specific product from its conception to launch, from order to delivery, and from raw materials into the hands of the customer.

Value stream mapping: the identification of all the specific activities (material and information flow) occurring during the production of a particular product or product family, usually represented pictorially in a value stream map. (*See* page 95 for icons.)

Visual control: the control of the workplace by the visual regulation of operations, performance goals, tool and parts placement, etc., so that a production process or other system can be understood at a glance. However, visual controls can appeal to any or all of the five senses.

Waste: basically, anything that adds cost or time without adding value. There are many different kinds of waste in manufacturing.

Work-in-process: material in the process of having value added to it—being converted into saleable goods.

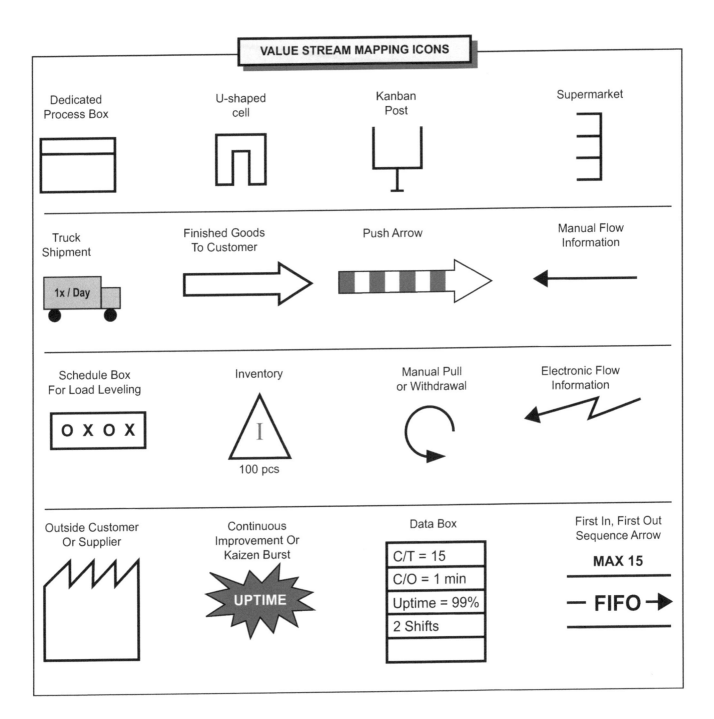

VALUE STREAM MAPPING ICONS

Dedicated Process Box

U-shaped cell

Kanban Post

Supermarket

Truck Shipment

1x / Day

Finished Goods To Customer

Push Arrow

Manual Flow Information

Schedule Box For Load Leveling

O X O X

Inventory

I

100 pcs

Manual Pull or Withdrawal

Electronic Flow Information

Outside Customer Or Supplier

Continuous Improvement Or Kaizen Burst

UPTIME

Data Box

| C/T = 15 |
| C/O = 1 min |
| Uptime = 99% |
| 2 Shifts |
| |

First In, First Out Sequence Arrow

MAX 15

— **FIFO**→

RECOMMENDED READING

Hirano, Hiroyuki. *5 Pillars of the Visual Workplace* (New York: Productivity Press, 1995)

Monden, Yasuhiro, *Toyota Production System: An Integrated Approach to Just-In-Time* (Norcross, Georgia: Engineering & Management Press)

Productivity Press Development Team. *Kaizen for the Shopfloor* (New York: Productivity Press, 2002)

Productivity Press Development Team. *5S for Operators* (New York: Productivity Press, 1996)

Sekine, Kenichi. *One-Piece Flow* (New York: Productivity Press, 1992)

Tapping, Don, et. al. *Value Stream Management* (New York: Productivity Press, 2002)

Tapping, Don, and Tom Shuker. *Value Stream Management for the Lean Office* (New York: Productivity Press, 2003)

Tapping, Don, et. al. *Value Stream Management Video Series* (New York: Productivity Press, 2001)

Womack, James, and Dan Jones. *Learning to See* (Massachusetts: The Lean Enterprise Institute, 1999)

INDEX

ABOUT THE AUTHORS

James Vatalaro

Jim launched his career in manufacturing in the U.S. aerospace industry over 20 years ago. He has spent more than 10 years implementing the principles of lean manufacturing by consulting in companies of all sizes and cultures around the globe. Although Jim has facilitated many hundreds of various kaizen and kaikaku events, he still considers himself a student of the Toyota Production System. He recognizes there is much more to discover about how to further improve production operations. Jim and his family reside in Earlton, NY.

Robert Taylor

Bob's manufacturing experience spans over 30 years. In the late 1980s, Bob learned lean manufacturing from the individuals who created it at Toyota. Since then, he has been transforming companies into lean enterprises. He is one of the first individuals to study and implement lean in his own manufacturing plants in the United States. Today, Bob is as devoted as ever to the practical implemention and teaching of lean practices worldwide. Bob and his wife reside in Narragansett, RI.

T - #0709 - 101024 - C0 - 280/208/9 - PB - 9781563272868 - Gloss Lamination